3-10-8

MARINE INVERTEBRATES

Comparative Physiology

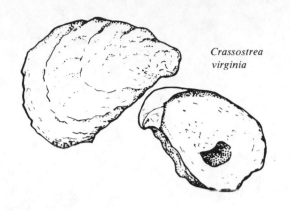

Crassostrea virginia

MARINE INVERTEBRATES

Comparative Physiology

CARL S. HAMMEN

Published for University of Rhode Island

by UNIVERSITY PRESS OF NEW ENGLAND

Hanover, New Hampshire and London, England 1980

UNIVERSITY PRESS OF NEW ENGLAND

Sponsoring Institutions
Brandeis University
Clark University
Dartmouth College
University of New Hampshire
University of Rhode Island
Tufts University
University of Vermont

Drawings by Susan Chandler Lum

Library of Congress Catalog Card Number 80–51505
International Standard Book Number 0–87451–188–7

Printed in the United States of America

Library of Congress Cataloging in Publication data
will be found on the last printed page of this book.

Contents

Figures

Tables

Preface

Marine invertebrates have aroused curiosity since the beginnings of science. They include such varied and fascinating animals as mollusks, crustaceans, echinoderms, polychaete worms, sponges, coelenterates, flatworms, nematodes, bryozoans, brachiopods, horseshoe crabs, and members of a dozen minor phyla. In fact, every phylum has some marine species, and several are exclusively marine.

Physiology is the science of life processes, their rates and underlying mechanisms. The purpose of this account is to describe processes precisely and to show how these vary from species to species, enabling each to cope with common problems of survival. Examples are taken from the major groups of marine invertebrates, those that have been sufficiently studied to make a coherent story. Calculations of rates are used to facilitate comparison between species or to show environmental effects.

The processes by which marine invertebrates sustain life are mostly basic cellular processes, such as osmotic adjustment and gaseous exchange. Many have simple body plans and low degrees of tissue specialization. In some groups, a particular organ system may be entirely lacking. With relatively large functional units and simple organization, some animals provide extremely valuable preparations for physiological research. The giant axon system of squids, the rudimentary brain of sea hares, and the large striated muscle fibers of barnacles are striking examples.

This book is intended to be read primarily by college seniors and first-year graduate students. It will also be useful to professional biologists working in related specialties. It can be used as a textbook for courses in comparative physiology, invertebrate physiology, and marine biology.

MARINE INVERTEBRATES

Comparative Physiology

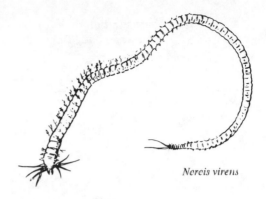

Nereis virens

CHAPTER 1 Osmotic balance

A jellyfish stranded on the beach dries out rapidly, but it cannot be rescued by placing it in freshwater, because marine invertebrates require water containing salts. Oceanic animals survive in water with a very limited range of salinity.

Salt content of sea water and body fluids

The principal salt of seawater is sodium chloride (NaCl), which makes up 30.8 g or 86 percent of the total of 35.8 g of salts in each liter. This concentration, found in most ocean waters of the world, is equivalent to $0.612M$ NaCl and to $1.10M$ sucrose or other nonelectrolyte. The other major cations of seawater are Mg^{++}, Ca^{++}, and K^+; the major anion is Cl^-, but $SO_4^=$ and Br^- are also present. Trace amounts of many other elements have been detected.

The body fluid of a jellyfish is very similar to seawater. In fact, the fluids of all marine invertebrates have nearly the same total osmotic concentration ("isosmotic") and nearly the same salt composition ("isoionic") as the waters they live in. For example, the coelomic fluid of the polychaete worm *Neanthes*

TABLE 1-1. Concentrations of salt ions and free amino acids (FAA) in tissues and body fluids (mM).

	Na	K	Ca	Mg	Cl	Total Salts	FAA
Neanthes succinea (polychaete)							
tissue	125	195	14	22	124	514	412
coelomic fluid	483	14	12	44	545	1098	9
seawater	479	9	13	37	569	1106	0
Crassostrea angulata (oyster)							
yellow muscle	188	158	9.2	—	162	517	441
seawater	492	15	8.1	—	673	1188	—
Carcinus maenas (crab)							
muscle	54	120	14	36	54	278	434
blood	468	12	35	47	524	1086	12
seawater	455	13	28	112	533	1141	0

(From Freel *et al.*, 1973; Bricteux-Gregoire *et al.*, 1964; Shaw, 1955)

succinea has a total salt concentration of 1098mM in water of 1106mM; the blood of the shore crab *Carcinus maenas*, 1086mM salts in a medium of 1141mM (Table 1-1).

Water inside the cells of these animals is also isosmotic, but often far from isoionic. In the case of *Neanthes*, the cells contain only one-half as much salt as expected, the bulk of the remaining osmotic effect being made up by free amino acids (FAA) in the cell water. A similar and even more striking example is seen in *Carcinus* muscle, where only one-fourth of dissolved substances consists of mineral salts. Organic compounds commonly make up 20 to 50 percent of osmotically active substances in tissues of marine invertebrates. Concentrations of both salts and organic compounds are adjusted as animals adapt to changes in salinity.

Not only is the total salt content within cells considerably less than that of seawater, but the proportions of the major ions vary markedly. Potassium ion (K^+) is accumulated, reaching

10 to 20 times the concentration in the water or the blood. Sodium (Na^+) is reduced to only 10 to 25 percent of the surrounding fluids. There is also a pronounced discrimination against Mg^{++} and $SO_4^=$ (Table 1-1). Some degree of ion regulation occurs in every species studied (10). Since no one knows precisely how the living cells handle these ions, we cloak our ignorance in the term "sodium pump." This refers to a hypothetical mechanism in the cell membrane that promotes extrusion of Na^+ and passive entry of K^+, so that on balance the interior is electrically negative but only slightly shifted from neutrality. The degree of effectiveness of this hypothetical mechanism is suggested by the steady-state ion ratios (K/Na) in cells of *Crassostrea angulata*, *Neanthes succinea*, and *Carcinus maenas:* 0.84, 1.56, and 2.22, respectively.

Responses to variations in salinity

Estuarine animals may experience large variations in salinity. When river discharge increases because of heavy rainfall or melting ice, the seawater near shore becomes more dilute. When prolonged droughts occur, areas partially isolated from the sea receive less freshwater and lose water by evaporation, thus becoming more saline. Animals respond to reduced salinity of their medium by taking in water; they respond to increased salinity by losing water (14). Weight changes due primarily to gain or loss of water are particularly conspicuous in soft-bodied, worm-like forms. Specimens of the polychaete *Nereis virens* tolerate abrupt transfer to seawater diluted to 25 percent of normal salinity. After eighty minutes, the worms weigh 60 percent more than their initial weight (Figure 1-1). When the seawater is nearer full-strength salinity (less dilute), the weight gain is smaller. And when NaCl is added to the medium, the worms diminish in weight. These worms behave as osmometers, gaining and losing water in accordance with the concentration gradient of their medium.

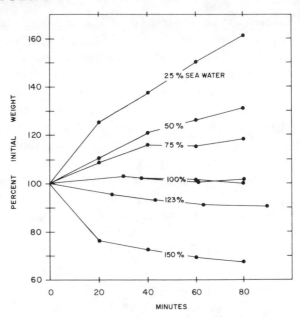

Figure 1-1. Weight changes of a polychaete *Nereis virens* on abrupt transfer to seawater of higher and lower salinity, acclimation salinity 31.1 g/l; percent of initial weight *vs.* time of immersion. Results of student exercises.

The amount of water normally present in an animal's body is characteristic of species; so is the amount of swelling from water influx that each can tolerate. For example, in the lugworm *Arenicola marina* in full-strength seawater, water makes up 87.5 percent of body weight and is distributed 67.5 percent in the coelomic fluid and 32.5 percent in tissues (7). Of the tissue water, 77 percent is in cells, 23 percent extracellular. Indefinite swelling by *Arenicola* is prevented by loss of solute.

In 50 percent seawater, nereid polychaetes reach their maximum swelling after four to ten hours, after which the animals slowly return to near their initial weight. The water intake ceases because they lose salts, and they eventually reach a new equilibrium. Both salts and water flow continuously across the cuticle and cell membranes, but the rate of water movement at first greatly exceeds the rate of salt movement and obscures it. There is a definite limit to osmotic adjustment by salt loss, imposed by the critical minimum salt content needed for operation of cellular processes. When the medium is diluted below this limit, animals must maintain ("regulate") internal salts above external concentration or perish.

Regulation implies either a decrease in permeability at low salinity or active uptake of salts to counterbalance salt loss, or both. By means of radioisotope tracers ^{22}Na and ^{36}Cl, rates of uptake can be determined with great precision. The estuarine polychaete *Nereis diversicolor* departs from osmotic conformity at low salinities, and active Na$^+$ uptake is demonstrated by the tracer method (18). The rate is maximal at 50mM Na$^+$ (10 percent seawater) and half-maximal at 10mM, suggesting that the "pump" behaves like an enzyme, becoming "saturated" at high salt concentrations. A mechanism for uptake of Cl$^-$ also operates when the worms are in diluted seawater. These active processes cease and passive exchange takes over at 50 percent seawater for Cl$^-$, 75 percent seawater for Na$^+$. The body wall of *Nereis diversicolor* also becomes less permeable to salts at very low salinities, which obscures the details of the pump mechanism but allows the animals to survive by conserving the necessary minimum salt content.

Species-specific adjustment of osmotic balance

Osmoregulatory powers vary with species within narrow taxonomic groups, such as families or genera. Most of the information about adjustment of osmotic balance has come from studies on polychaetes, mollusks, and crustaceans. In adjusting to reduced salinity, mollusks do not take in large amounts of water, as polychaetes do. For example, the blood of a littoral gastropod, *Littorina saxatilis*, becomes isosmotic with 60 percent seawater in one hour, but tissue wet weight increases only 11.5 percent. An increase in chloride in the medium indicates that the snails release salt, thus avoiding a large weight gain due to influx of water (2). When a chiton, *Acanthochitona discrepans*, is placed in 50 percent and 25 percent seawater, the total osmotic concentrations and the NaCl concentrations of the blood become equal to these mediums within five hours (8). Bivalves can delay adjustment to altered salinity by tightly closing the valves for various periods, depending on species and conditions. For example, the blood of *Glycymeris glycymeris* remains hyperosmotic to 25 percent seawater for five days, but becomes isosmotic after a week or more. Both the chiton and the bivalve maintain blood K^+ in diluted mediums while other ions diminish, suggesting that K^+ is more important to normal function than the others. In some bivalves, such as *Mytilus edulis*, the free amino acid concentration of tissues is reduced by more than one-half on acclimation to 50 percent seawater, suggesting that this species prefers amino acid loss to salt loss.

Crustaceans resist dilution by means of impermeability of the rigid exoskeleton and indeed most of the body surface. Salt and water exchange occur in crustaceans mainly across the gills, which must be permeable enough to allow for exchange of gases. All degrees of osmoregulation are found among the members of this diverse group, from none to complete, the degrees correlated with habitat. For example, the spider crab, *Maia*, entirely marine, has blood isosmotic with the medium down to 60 percent seawater and fails to survive lower salinities; the shore crab, *Carcinus*, an estuarine animal, has blood hyperosmotic to the medium at salinities from 100 percent seawater to 10 percent seawater, but isosmotic at higher-than-normal salinities; and the fiddler crab, *Uca*, which is semiterrestrial, maintains its blood osmotic pressure within relatively narrow limits, responding in only a minor way to the salinity of the medium (14).

Mechanisms of osmoregulation

Water and salt balance is continuously adjusted in body tissues as well as in the blood. The water content of tissues of *Carcinus maenas* does not vary significantly with osmotic stress (18). The blood acts as an osmotic buffer, protecting the tissues from direct exposure to large variations in the salinity of the external milieu. Cell membranes of the cells in the gill tissue must determine which ions are lost and how rapidly they escape. The membranes are more than simple barriers, for they contain machinery for actively transporting salts and organic substances; there is evidence that even their structure is dynamic rather than static.

The search for mechanisms of osmoregulation requires close examination of the earliest effects of exposure to altered mediums. Salt loss begins immediately upon immersion of crabs in diluted seawater, then gradually slows down as equilibrium is approached. This process has been studied by measuring the electrical conductivity of the medium, which increased as salts left the animal and increased the salinity of the water. The results were expressed as rate constants (k), the values of the exponents in exponential equations selected to fit the data. These values fell in the range 0.26–0.35 for *Carcinus maenas;* the related half-times to equilibrium were 2.0–2.8 hours. These experiments gave the additional information that crabs were able to change their permeability to salts in such a manner that permeability was minimal at a salinity of 20 g/l, an intermediate level likely to be found in estuaries (19).

Another early effect of placing *Carcinus maenas* in dilute seawater is an increase in oxygen consumption, signaling that cellular oxidation is accelerated, and more energy is made available for cellular processes. One way to avoid dilution of cytoplasm would be to have a kidney which pumped out a watery urine; this would require extra energy, resulting in greater oxygen consumption. In fact, there was a fourfold increase in urine production by the antennal gland in 50 percent seawater

(3). However, the increases in respiration usually observed are much greater than the calculated energy required for osmotic work (10).

An increase in ammonia excretion by *Carcinus maenas* has been detected within four hours (9). This implies that substances containing nitrogen were being degraded. The tissues of *Carcinus* contain large quantities of free amino acids (FAA), estimated at 40 to 60 percent of the total of osmotically active substances. The tissue FAA concentrations diminish during acclimation; since there was no increase in FAA loss in dilute seawater, it is probable that the loss during acclimation occurs through oxidative deamination. This cellular process could account for increases in both ammonia excretion and oxygen consumption.

In the polychaete, *Nereis virens*, tissue FAA make up 22 to 24 percent of the total of osmotically active substances and remain at that percentage while concentrations of all substances diminish during acclimation to 50 percent seawater. No FAA are detected in samples of the medium. Ammonia excretion is greater in dilute seawater even during the first thirty minutes of acclimation; this excretion rate continues to be high for one to five hours, while the excretion rate in 100 percent seawater diminishes to less than one-third of the initial rate (Figure 1-2). This strengthens the hypothesis that oxidation of FAA is a means of disposing of superfluous osmotic substances while simultaneously obtaining energy to adjust salt and water balance. A review of much research on this problem concluded that "the mechanisms involve at least an increased catabolic rate of amino acids" (12).

The reverse process, raising the internal osmotic content in response to increased salinity, is largely a matter of water loss. At salinities above 75 percent seawater, the exchange of both Na^+ and Cl^- in *Nereis diversicolor* was passive, or directly proportional to external concentrations (18). The free amino acid pool can be refilled on exposure to higher salinities. Taurine is a sulfonic amino acid that does not occur in the structure of

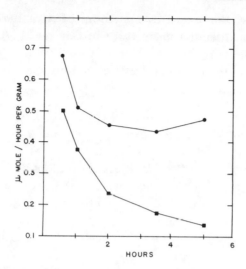

Figure 1-2. Rates of ammonia release by a polychaete *Nereis virens* in seawater (lower curve), and 50 percent seawater (upper curve); μ *mole*/hour per gram *vs.* time of immersion.

proteins, as do the twenty common carboxylic amino acids. The only known function of taurine is that of osmotic effector. The bivalve *Mya arenaria* loses about half of the taurine content of its tissues in dilute seawater; it restores the taurine level when returned to full-strength seawater (1).

Differences between groups indicate differences in permeability of cell membranes. In general, mollusks have the most permeable membranes; loss of salt and FAA are rapid. Polychaetes release salts slowly and FAA not at all, but they are very permeable to water, and therefore undergo large volume changes. Crustaceans are largely impermeable, except for the gills, where salts are released fairly rapidly. They degrade most FAA to ammonia and release only small quantities unchanged.

Tissue concentrations of free amino acids are adjusted to changing salinity—the question is by what means. One hypothesis is that glutamate dehydrogenase (EC 1.4.1.2) activity may be directly affected by cation concentrations (16). ("EC" stands for Enzyme Commission, the authors of a taxonomic scheme for enzymes, explained more fully in Chapter 4; see page 37.) Glutamate dehydrogenase (GDH) is one of the few enzymes known to catalyze a net synthesis of an amino acid:

$$\text{2-oxoglutarate} + NH_3 + NADH \rightleftharpoons \text{L-glutamate} + H_2O + NAD^+$$

With an enzyme preparation from a lobster *Homarus vulgaris*, activity in the glutamate-forming direction was shown to be stimulated by ammonium salts, and Cl^- was found to be "a good activating anion." In fact, NaCl at $400mM$ produced great increases of reaction velocity with GDH from both gill and muscle. This suggests that glutamate synthesis could be controlled directly by the salt content of the animal's environment. The supply of glutamate can influence concentrations of alanine, aspartate, and other amino acids by means of transamination reactions, which have been found in many species. The GDH hypothesis thus provides a possible way that salinity can directly determine FAA levels without a need in the animal for elaborate sensing and control mechanisms.

The GDH hypothesis is less useful for explaining reduction in FAA levels and increased ammonia excretion during adaptation to reduced salinity, because activities in the deaminating direction are undetectable in some polychaetes and mollusks, and amount to only 1 to 10 percent of activities in the glutamate-forming direction in crustaceans (4). Alternate pathways that may be more useful include a purine nucleotide cycle and a serine cycle. These schemes employ transamination reactions and are coordinated with portions of carbohydrate metabolism, the final steps catalyzed by:

adenylic acid deaminase (EC 3.5.4.6), $AMP + H_2O \rightleftharpoons$
$IMP + NH_3$

and serine deaminase (EC 4.2.1.13),

$\text{L-serine} + H_2O \rightleftharpoons \text{pyruvate} + NH_3 + H_2O$

These deaminases are widely distributed; the serine enzyme is more active at lower salt concentrations, which is appropriate for reducing FAA content. Convincing evidence in favor of one of these pathways over the others is not yet available. However, it is clear that osmotic adjustment and nitrogen excretion (Chapter 7) are considered separately only for convenience of study and are actually closely integrated processes.

Balannus eburneus

CHAPTER 2 Gaseous exchange

From the moment that any aquatic animal is isolated in a closed vessel it begins to deplete the water of dissolved oxygen and contribute a roughly equivalent amount of carbon dioxide to it. Even while motionless the animal changes the concentration of both gases markedly in a few minutes. This gaseous exchange is sometimes called "respiration," because O_2 and CO_2 are major reagent and product, respectively, of cellular respiration.

Most small animals, and some larger ones such as jellyfishes, depend on simple diffusion across their surfaces for gaseous exchange with the medium. Larger forms have gills, respiratory tentacles, or other specialized structures that increase the surface available for diffusion; some carry on ventilatory movements to promote exchange. Many animals possess respiratory pigments, such as hemocyanin (Chapter 3), which combine loosely with oxygen at the gills, circulate through vessels and spaces, and release O_2 in other tissues and organs.

The solubility of gases

The amount of O_2 reaching a respiratory surface depends on the solubility of O_2 and on its partial pressure in the air above the medium. For example, the fraction of O_2 in air is 0.2095 and its solubility in pure water at $0°$ C is 49.1 ml/l; therefore, the actual concentration of O_2 in the cold water is $0.2095 \times 49.1 =$ 10.29 ml/l (7). The average O_2 content of the sea has been estimated at 2.5 ml/l. There are several reasons for this relatively low value: solubility of gases in water diminishes with increase in temperature (Table 2-1), and the average temperature of the entire earth's surface is not $0°$ C, but about $12°$ C. The presence of salts also reduces the solubility of O_2. For example, oxygen is soluble in standard seawater only to the extent of 78.5 of its solubility in pure water. A third factor that reduces the average O_2 content is that much of the mass of water at great depths lacks contact with the atmosphere.

The concentration of CO_2 in water would be enormous if it were present as a pure gas, but CO_2 is only a minute fraction of air: 0.00032. Thus, the actual concentration of CO_2 in pure water at $0°$ C is: $0.00032 \times 1713 = 0.55$ ml/l, or only 1/20 of the O_2 in solution. In seawater, dissolved CO_2 gas is in equilibrium with the salt ions HCO_3^- and $CO_3^=$. The proportions vary with acidity. At pH 8.0, a common value for seawater, most of the total CO_2 is in the form of bicarbonate ion and there is practically no CO_2 gas (7).

TABLE 2-1. Solubility of gases at 1 atm in pure water at various temperatures (ml/l).

$°C$	O_2	CO_2	N_2
0	49.1	1713	23.3
15	35.0	1010	17.0
20	31.0	901	15.4
25	28.0	759	14.5
30	36.6	625	14.1

Rates of oxygen consumption

In practice, the amount of dissolved O_2 usually is kept constant during measurement of O_2 utilization, by keeping the water in equilibrium with the air above it and measuring for relatively short periods. Rates of O_2 consumption are commonly determined by analyzing samples chemically before and after an animal has been confined in a closed vessel (Winkler method), or by recording pressure changes in a manometer at regular intervals (Warburg respirometry). In the Warburg respirometry method, the vessel contains alkali in a separate chamber to absorb the liberated carbon dioxide according to the following reaction:

$$2KOH + CO_2 \rightarrow K_2CO_3 + H_2O$$

As the dissolved salt has virtually no volume compared to the gas, a fall in pressure is recorded by the manometer. Calibration of the Warburg glassware allows calculation of the actual volume of oxygen used (9). Readings can be taken at ten-minute intervals for an hour or more. The usual finding is a linear increase in total O_2 consumed with time, just as one would expect of an essentially continuous process.

Rates of O_2 consumption are affected by many factors, and they are expressed in various ways. Larger animals generally consume more than smaller individuals of the same species, so it is useful to place rates on an equal-weight basis. For example, three barnacles of the species *Balanus eburneus*, weighing 0.800 g, consumed 0.053 ml O_2 in 40 minutes at 20° C. The hourly rate is (60 min/40 min) × (0.053 ml O_2) = 0.080 ml/hr; the weight-specific rate is (0.080 ml/hr) ÷ (0.800 g) = 0.100 ml/hr per g.

For comparison with other processes, a conversion of volume of O_2 consumed to moles of O_2 consumed can be useful. The gram-molecular volume at standard conditions is 22.4 l. The rate then is $(100 \times 10^{-6} l) \div (22.4 l/mole) = 4.46 \mu$ moles O_2/hr per g.

The effect of size on the O_2 consumption of an animal is

partially compensated for, but not totally eliminated by calculating weight-specific rates of consumption. An individual weighing twice as much as another does not consume twice as much, but considerably less than twice. For many species the individual rates are proportional to the 2/3 power of mean weight: $y = kW^{2/3}$, where y is μl O_2/hr, W is weight in mg, and k is some empirical constant of proportion. For example, if a barnacle of 266 mg consumes 26.6μl O_2/hr, we might expect an animal of 532 mg to consume 53.2μl O_2/hr; but we find the actual rate is 42.0μl/hr. In this case: $y = kW^{2/3}$, $26.6 = k(266)^{2/3} = 41.36k$, and $k = 0.643$. For the larger specimen: $42.0 = k(532)^{2/3} = 65.65k$, and $k = 0.640$. The close agreement in values of k indicates that the exponent has been correctly chosen. This relationship between respiration and weight is called "the Surface Law," because the surface of any solid figure varies as the square of linear dimension ($S \propto L^2$); its volume varies as the cube ($V \propto L^3$). Therefore, its surface is proportional to the 2/3 power of its volume ($L \propto S^{1/2} \propto V^{1/3}$, $S \propto V^{2/3}$). With density constant, weight and volume are directly proportional. The Surface Law states that respiratory rate is proportional to the 2/3 power of weight. An illustration that the rule applies to three species of barnacles is shown on the accompanying graph (Figure 2–1).

For a given concentration gradient, the rate of gaseous diffusion across a surface depends on the amount of surface available; so in a sense, animals appear to have adjusted their O_2 utilization to the rate at which their O_2 can be supplied and their CO_2 eliminated. Another explanation of the Surface Law is that heat production is an inevitable condition of metabolism; also, heat loss by radiation is proportional to the surface available. For example, heat production by the whelk, *Nucella lapillus*, is 0.080 cal/hr per mg N; O_2 consumption is 0.0168 ml/hr per mg N (3). This rate of heat production is equivalent to that expected from combustion of endogenous amino acids by the O_2 consumed. As marine invertebrates lack mechanisms of temperature control, they may have adjusted their respiratory rate in

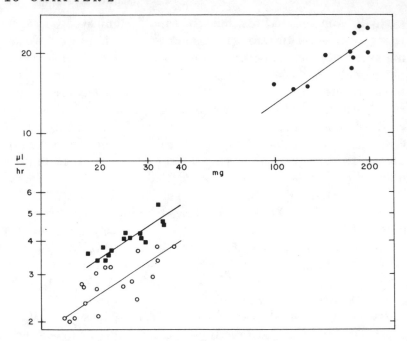

Figure 2-1. The Surface Law. Oxygen consumption rates (μl/hr) of three species of barnacles increase as the 2/3 power of total body weight without shell. Equation of lower left line is $y = 0.345W^{2/3}$. Others have the same form (4).

accordance with their surface area to permit a steady, low rate of heat loss to the environment.

Oxygen consumption and tissue weight

Measurements are sometimes expressed in terms of unit weight of N because animals vary greatly in their fraction of meta-bolically inert material, such as their shells, and also in the water content of their tissues (Tables 2-2 and 2-3). Since only tissues respire, a refinement is effected by expressing O_2 consumption per unit weight of tissue rather than unit weight of whole animal. A mussel weighing 4 g has twice as much living tissue

TABLE 2-2. Proportions by weight of parts of shelled marine invertebrates (percent of total).

Species		shell	fluid	tissue
Crassostrea virginica	oyster	73.0	19.0	8.0
Modiolus demissus	mussel	52.9	30.5	16.6
Monodonta sagittifera	snail	69.4	7.3	23.4
Portunus pelagicus	red crab	53.0	- - - -	47.0
Chthamalus depressus	barnacle	84.8	6.0	9.2
Terebratulina septentrionalis	brachiopod	40.8	44.7	14.5
Lingula reevi	brachiopod	19.5	18.7	61.8*

*Including pedicle

TABLE 2-3. Ratios between components of tissues.

Species		$\dfrac{\text{dry weight}}{\text{total}}$	$\dfrac{\text{protein}}{\text{dry}}$
Pocillopora capitata	coral	0.084	0.330
Mnemiopsis leidyi	ctenophore	0.034	0.074
Chaetopterus variopedatus	polychaete	0.144	0.580
Asterias forbesi	sea star	0.330	0.342
Crassostrea virginica	oyster	0.146	0.538
Loligo pealei	squid	0.207	0.740
Penaeus aztecus	shrimp	0.238	0.900
Portunus pelagicus	crab	0.140	0.736

(proportionately) as an oyster weighing 100g (Table 2–2). On
the basis of whole body weight, the O_2 consumption of the
oyster is 0.039 ± 0.010 ml/hr per g, and that of the mussel,
0.029 ± 0.005 ml/hr per g, at $25°$ C (5). The oyster rate appears
to be only one-third greater, a relatively insignificant amount
given the inherent variation. However, on a tissue basis these
rates become 0.488 and 0.175 ml/hr per g tissue, a difference of
2.8-fold.

Marine invertebrates generally have greater tissue water content than animals from other habitats (Table 2–3). This means
that the solid materials making up the respiratory apparatus of
cells are less abundant, and overall metabolism should be slower.
In fact, the tissue respiratory rate of the oyster, 0.488 ml

O_2/hr per g, seems small when compared to the 0.600 ml O_2/hr per g for a butterfly at rest (8). On a dry weight basis, however, the oyster rate becomes 0.488/0.15 = 3.25 ml/hr per g dry, and the butterfly rate is less than half of that, or 1.50 ml/hr per g dry, since insects contain only 48 to 61 percent water.

The dry material of tissue consists of a great number of compounds, principally salts and the major food substances—proteins, fats, and sugars. Some tissues contain large amounts of storage compounds—fats or polysaccharides. These play little part in oxidative metabolism. On the other hand, the supply of enzymes is critical, and all enzymes are proteins. Of the dry material, 41 to 74 percent is protein. The fraction of proteins made up by the element nitrogen is relatively constant, 12 to 19 percent, with an average of 16 percent. Therefore a measurement of nitrogen content will give the magnitude of active or potentially active enzymatic machinery. When processes such as O_2 consumption or heat production are evaluated according to unit weight of nitrogen, much of the variation in values owing to differences in amount of metabolically inert materials disappears. The same rate of O_2 consumption, expressed in various manners for the oyster, is as follows:

Rates per g of:	Fraction of preceding weight	ml/hr	μmoles/hr
whole animal		0.039	1.74
tissue only	0.08	0.488	21.8
dry weight	0.15	3.25	145.5
protein	0.41	7.90	397.0
nitrogen	0.16	49.3	2200.0

The last rate listed is equivalent to 2.20μ moles/hr per mg N at 25.0° C. The rate of *Nucella* was 0.75μ mole/hr per mg N at 20.4° C. The sedentary oyster, therefore, has a higher respiratory rate than the gastropod; partly due to the higher temperature, and partly a consequence of pumping activity, moving large volumes of water through the valves.

Difficulties arise in attempting to compare rates when the

amount of biomass is expressed in various ways, such as dry weight, wet weight, and weight of nitrogen. Table 2-3 shows ratios between weights of various components of living tissue. The differences between species are greatest in weight of tissue per unit whole-body weight (Table 2-2), so this is an important correction to make when presenting rates of O_2 consumption or other processes. The other ratios show much less variation and can be neglected, except when comparing marine invertebrates with terrestrial or freshwater forms.

The respiratory quotient

Consumption of O_2 is always accompanied by production of CO_2. The rate of CO_2 production, therefore, is also a kind of respiratory rate. The ratio between the two rates, CO_2/O_2, is called the respiratory quotient (RQ). It gives an indication of the kinds and proportions of food substances being oxidized. General equations representative of carbohydrates, lipids, and proteins are:

$C_{12}H_{22}O_{11} + 12O_2 \rightarrow 12CO_2 + 11H_2O$
Sucrose
   RQ = 12/12 = 1.0

$C_{57}H_{110}O_6 + 81\frac{1}{2}O_2 \rightarrow 57CO_2 + 55H_2O$
Tristearin
   RQ = 57/81.5 = 0.70

$C_{3032}H_{4816}O_{872}N_{780}S_8F_4 + 3800O_2 \rightarrow 3032CO_2 + 2408H_2O +$
   $780N + 8S + 4Fe$
Hemoglobin
   RQ = 3032/3800 = 0.80

These equations actually indicate the overall effect of burning foods in a calorimeter, but the proportions of reagents and products are the same when the substances are degraded through enzyme-catalyzed reaction series within animal cells. Measurements of gaseous exchange and heat production of experimental animals, such as guinea pigs, and of human subjects on defined diets, have proved that the RQ does indeed vary predictably with the kinds and amounts of foods consumed (1). Although good data are lacking, the relation is probably valid for marine invertebrates.

Carbon dioxide production

Measurement of CO_2 produced is more subject to error than measurement of O_2 consumed, because tissue respiration of marine invertebrates must be determined with the tissue in natural seawater or mediums of similar composition, and such mediums are complex with respect to carbon dioxide. A major complicating factor is that the equilibrium between forms of CO_2 (gas in the air, gas dissolved in the water, bicarbonate ion and carbonate ion) is influenced by pH, which may change during the course of an experiment. The safest course is to keep incubation times brief, and to use more than one method, in the hope that one will verify another. The CO_2 produced may be removed from the air by trapping it in a standard amount of alkali; the excess alkali is then titrated with standard acid. Another method is to measure net gaseous exchange in one set of flasks, and O_2 consumption in a duplicate set. The difference between the two sets represents CO_2 production, when appropriate corrections are made for the differences in solubility. This "paired flask" method has the disadvantage that one group of animals or tissue respires under an atmosphere gradually increasing in CO_2 concentration, while the other respires under an atmosphere where the CO_2 concentration is constant at near zero, because of its continuous removal by alkali. The O_2 consumption is assumed to be identical under the two sets of conditions, but this is a risky assumption, because CO_2 concentration is known to affect respiration rates of some animals.

A better method is to place reagents in isolated compartments in a flask, measure net exchange for a short period, then mix the reagents to form KOH without opening the flask, and determine O_2 uptake over the same interval (11). The chemical reactions representing this clever experimental trick are:

$$K_4Fe(CN)_6 + KMnO_4 \rightarrow MnO_2 + K_2O + K_3Fe(CN)_6$$
Ferrocyanide Permanganate

$$K_2O + H_2O \rightarrow 2KOH$$

The difference between net gaseous exchange and O_2 consumption is the CO_2 production. Rates of each process in a series of determinations are best presented in a single graph in which the RQ is shown by the slope of the best line through the experimental points, as in Figure 2–2.

Animal diet and the respiratory quotient

If an animal's nutritional state is unknown, determination of RQ can give a clue. An RQ near 1.0 indicates utilization of carbohydrate, values near 0.7 mean lipids are being used, and the RQ corresponding to a mixed diet is 0.80–0.85. The mean RQ of intact oysters, *Crassostrea virginica*, determined by sampling the medium, was 0.833 (2). A large range of variation, 0.51–1.44, was probably caused by some oysters retaining CO_2, and others releasing it in bursts. Oysters enter an anaerobic state when their valves are tightly closed, and metabolic acids dissolve the calcium carbonate of the shell (Chapter 7), releasing extra CO_2. Nitrogen excretion rates are low, indicating only about 2 percent protein in the diet of *C. virginica*. Therefore, the remaining 98 percent would consist of RQ = 0.833 = 1.00x + 0.70(1 - x), where the average RQ consists of a fraction of carbohydrate (x) and a fraction of lipid (1 - x); x = 44.3% carbohydrate; and 1 - x = 55.7% lipids. Oysters feed on plankton, including diatoms which are rich in lipids, up to 25 percent of organic materials. If the lipids of diatoms were used completely, and the carbohydrates only in part, due to differences in digestive enzymes (Chapter 4), then our calculation of 55.7 percent lipids actually used in the diet could be correct.

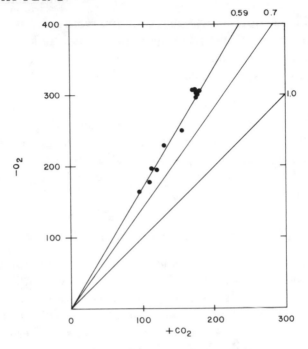

Figure 2-2. Respiratory quotient. Oxygen consumption and carbon dioxide production rates (μl/hr. per g) of a marine triclad flatworm *Bdelloura candida* were determined by producing KOH in Warburg flask after obtaining net gaseous exchange. Slope of best line through 11 points indicates RQ of 0.59.

Respiratory quotient and carbon dioxide fixation

Respiratory quotients outside the range 0.7–1.0 usually indicate defects in techniques used to measure CO_2 production. However, unusually low RQ can sometimes be accounted for by the ability of an animal to withhold and use a fraction of metabolic CO_2 for synthetic purposes. For example, a marine triclad flatworm, *Bdelloura candida*, commonly found on gills of horseshoe crabs, had oxygen consumption of $15.0\,\mu$ moles/hr per gram and CO_2 production of $9.6\,\mu$ moles/hr per g, giving RQ = 0.64 after four days without food (6).

Bdelloura, with an average initial weight of 5.3 mg, lost weight at the rate of 0.13 mg/day and released ammonia at the rate of 0.24 μ mole/day. At a tissue water content of 80 percent, the loss of weight amounts to 26 μg dry material/day, of which 0.3 μg is salt. The loss of ammonia is equivalent to 3.34 μg N/day. If the animal's protein is 16.0 percent nitrogen, a reasonable average value, then it was oxidizing its own protein at the rate of 20.9 μg/day. This accounts for 81 percent of the weight loss during starvation. The expected RQ for protein oxidation is 0.80, and the RQ found was 0.64. CO_2 fixation, measured with the use of radioactive bicarbonate, amounted to 2.36 μ moles/hr per g. This quantity added to the CO_2 produced gives corrected RQ = (9.6 + 2.36)/15.0 = 0.80, the expected value.

Heterotrophic carbon dioxide fixation, which depends on chemical energy to convert 3-carbon acids to 4-carbon acids, is universal in the animal kingdom. Rates of CO_2 fixation vary widely, but at least in the case of *Bdelloura*, they can account for the reduction of CO_2 output that causes some respiratory quotients to fall below the normal range.

Standard metabolism

Activity leads to increases in gaseous exchange, and all successful methods must either restrain the animal or permit only minimum activity. Even in quiescent animals, there is often beating of cirri or cilia, as in the pumping activity of the oyster, mentioned above. The O_2 consumption at the minima of activity is sometimes called "standard metabolism." Metabolism is the aggregate of all chemical reactions that occur in living tissue, and under most circumstances the respiratory rate is a fairly good indicator of its magnitude.

Degrees of conformity

Most species of marine invertebrates adjust their oxygen consumption rate ($\bar{V}$) to the partial pressure of oxygen (P) in

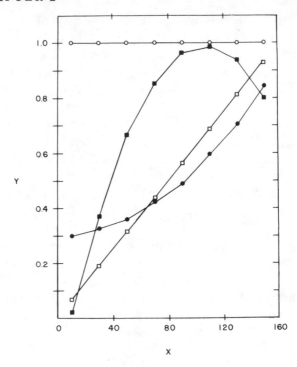

Figure 2–3. Respiratory conformity and regulation. Rates of oxygen consumption in real animals increase with partial pressure of oxygen in medium, and departure from strict conformity is described by quadratic equations (see p. 25). Y is fraction of maximal oxygen uptake; x is partial pressure of oxygen.

their immediate environment. This is known as respiratory conformity; some degree of conformity is the usual rule. For example, three specimens of the sea anemone, *Actinia equina*, consumed 1.4 ml O_2/hr at 0.289 atm partial pressure, but much less, 0.5 ml/hr at a lower pressure, 0.072 atm (8). An animal that respires at the same rate regardless of O_2 tension is called a respiratory regulator. However, there appear to be no complete conformers or regulators. Exact description of a species according to the magnitude of its departure from strict respiratory conformity is accomplished by the simple quadratic equation (10):

$$y = B_0 + B_1 x + B_2 x^2 \quad \text{where} \quad y = \bar{V} \quad \text{and} \quad x = P$$

The values of the parameters, B_0, B_1, and B_2 are chosen to make the calculated curve fit the actual graph of $\bar{V}$ vs. P. In practice, y (or $\bar{V}$) is a relative or normalized oxygen consumption rate, expressed as a fraction of the maximum observed; x is expressed as mm of mercury.

For example, the response of a bivalve (Figure 2-3) is represented by: $y = -0.19 + 22.2 \times 10^{-3}x - 10.5 \times 10^{-5}x^2$, and the ophiuroid respiratory rate varies with oxygen tension according to: $y = 0.30 + 2.38 \times 10^{-5}x^2$. The bivalve tends to regulate P above 80mm Hg, and the ophiuroid at P below 80mm Hg. Deviations from conformity are precisely described by the appropriate equations. Now the search for underlying control mechanisms can begin.

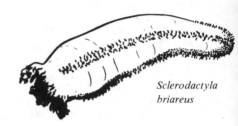

Sclerodactyla
briareus

CHAPTER 3 Oxygen transport

The blue blood of lobsters and the red blood of many poly-chaete worms owe their color to special pigments used to transport oxygen. Hemocyanin, the blue pigment, and hemo-globin, the red pigment, are proteins that combine with oxygen in the gills, where oxygen is readily withdrawn from the surrounding water. In coelomate animals, this complex is carried by a blood-vascular system to muscles, viscera, and other internal tissues, where oxygen is released to be used in tissue respiration. Such systems generally include one or more con-tractile segments ("hearts") that produce a pressure gradient. The oxygen-depleted pigment then returns to the gills, com-pleting a continuous circulatory movement.

Distribution of hemocyanin and hemoglobin

Hemocyanin is dissolved in the blood of gastropod, am-phineuran, and cephalopod mollusks, many large crustaceans, and the horseshoe crab *Limulus*. Because of the size of these animals and their dependence on the blood for oxygen, hemo-cyanin is probably the most important invertebrate respiratory pigment.

Hemoglobins are more widely distributed, occurring in many polychaetes, some echinoderms, some bivalve mollusks, and in smaller crustaceans such as barnacles, branchiopods, copepods, and ostracods. Of the major groups, only sponges and coelenterates seem to lack species with hemoglobin or hemocyanin.

Properties of hemocyanin

Hemocyanins are large protein molecules with molecular weights from 8.25×10^5 in the lobster *Homarus* to 8.8×10^6 in the whelk *Busycon*. They contain copper, 0.16 to 0.19 percent by weight, corresponding to 20 to 200 copper atoms per molecule. One oxygen molecule is bound for each pair of Cu^{++} atoms.

When combined with oxygen, the pigment is blue; when deoxygenated, it is colorless. Maxima of absorption are at 335–350 nm and 580–585 nm for the pure compound. The oxygenated form absorbs more light at all wavelengths. The absorption spectrum of blood of *Homarus americanus* is shown in Figure 3-1. In this case, the peak in the visible region occurs at 560 nm.

The oxygen-carrying capacity of bloods containing hemocyanins ranges from 7.0 ml O_2 per liter in *Limulus* to 42.7 ml/l in the squid *Loligo*, with an average of about 25 ml/l. This can be compared to seawater or body fluid without pigment, which contains, when fully aerated, about 6 ml O_2 per liter at 15° C. Thus the increase due to hemocyanin can be small or as much as seven-fold. The actual blood concentration of hemocyanin is 40 g/l in *Busycon* and 80 g/l in *Loligo*, or $4.5 \times 10^{-6} M$ and $21.0 \times 10^{-6} M$. Such concentrations have little effect on the total osmotic strength of the blood.

Oxygen dissociation curve

An important property of whole blood is the relation between degree of saturation with O_2 and the partial pressure of the gas

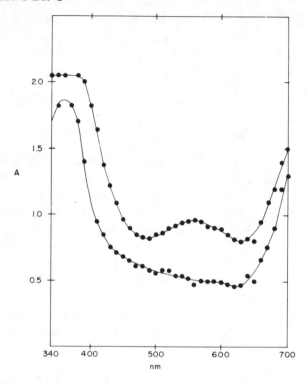

Figure 3-1. Absorption spectra of hemocyanin from lobster *Homarus americanus*, with and without oxygen. Results of class exercise.

in contact with it. This relation is established in the following manner: blood is spread in a thin film on the inner surface of a vessel called a tonometer, which can be emptied and filled in turn with gas mixtures containing various fractions of oxygen. The tonometer fits in the cell compartment of a photometer where the light absorbance at a peak, for example 580 nm, is determined for fully saturated and partially saturated samples. The ratios of these absorbances are plotted on a graph against partial pressure of O_2 to give a characteristic figure called an oxygen-dissociation or oxygen-equilibrium curve (3).

The hemocyanin of the blue crab *Callinectes sapidus* (11) is described by the curves of Figure 3-2. The maximum O_2 tension used experimentally is that of air: 0.21×760 mm Hg = 160 mm Hg, but this pigment is fully saturated at much lower tensions, so the scale goes only to 40 mm Hg. All the action of loading and unloading, occurs at comparatively low oxygen tensions in these animals. Half saturation (P_{50}) is at 4 mm Hg in *Callinectes*. P_{50} is often used to characterize an oxygen carrier, as an indicator of the midpoint of its equilibrium curve.

In four species of chitons, for example, the hemocyanins have P_{50} of 20-25 mm Hg and P_{95} of 60 mm Hg (7). Two species required slightly higher O_2 pressures to produce half saturation at higher hydrogen ion concentration $[H^+]$. This is known as the Bohr effect (1). A third species showed no effect of $[H^+]$ and a fourth actually had less affinity for O_2 as $[H^+]$ decreased.

In most species, the properties of hemocyanin correlate well with the O_2 needs of the animal and O_2 concentrations encountered in its habitat (8).

The properties of hemocyanin are related to habitat in a tropical population of the blue crab *Callinectes sapidus* (11). The coagulated blood of the crab was filtered, centrifuged, placed in a tonometer, and deoxygenated with a vacuum pump. Then air was admitted into the vessel in increments, and the absorbance (A) was measured with a spectrophotometer at 580 nm wavelength. Air was assumed to be 0.2094 oxygen, and A at each increment was assumed proportional to A at full saturation.

P_{50} was 2.0 mm Hg at 23° C. and 5.6 mm Hg at 28° C. A pronounced Bohr effect is shown by the oxygen-affinity curves (Figure 3-2). Thus release of O_2 in the tissues of *Callinectes* is promoted by both temperature and CO_2. These are properties of survival value to a respiratory conformer in tropical areas, where high temperatures and salinities reduce the O_2 content of the water.

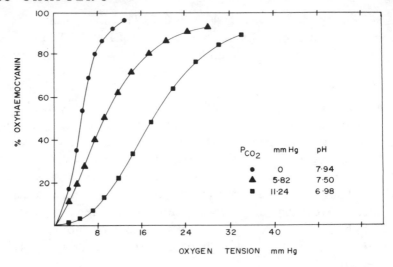

Figure 3-2. Oxygen equilibrium curves of blue crab *Callinectes* hemocyanin. Succeeding curves, left to right, represent results at pH 7.94, 7.50, and 6.98, demonstrating Bohr effect. (From Young, 1972).

Properties of hemoglobin

Hemoglobin consists of an iron-porphyrin, heme, combined with a protein, globin. There are four heme groups (Figure 3-3), each with one atom of ferrous iron (Fe^{++}) in a molecule of hemoglobin. The iron atoms in the heme portion of the molecule are the binding sites for O_2. Porphyrin compounds are widely distributed in nature. In addition to heme, they include the cytochromes, the chlorophylls, and enzymes such as catalase and peroxidase. The great variation in molecular weight of hemoglobins from different species (18,200–3,000,000) is due to the size of the globin portion (Table 3-1).

The smaller hemoglobins are often held in circulating cells rather than simply dissolved. This eliminates their osmotic effect and prevents loss by filtration through cell membranes of the vascular system. The hemoglobin of *Glycera*, mol wt 18,200, occurs in coelomocytes; if it were in solution, a concentration

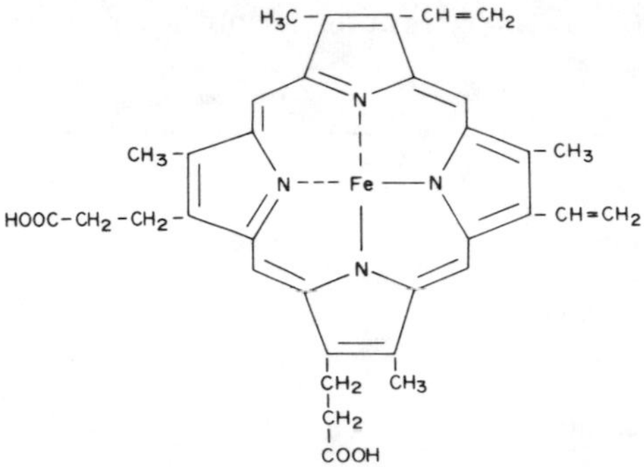

Figure 3-3. Structure of heme, the iron-porphyrin compound that holds oxygen in hemoglobin and myoglobin.

of 140 g/l would have an osmotic effect of an organic solute at 8 mM.

Porphyrin compounds and light absorption

The oxygen-carrying ability of hemoglobin and other blood pigments is measured by means of differences in light absorption between oxygenated and unoxygenated forms. All porphyrins have a strong absorption at 400–420 nm, a region in the near ultraviolet known as the Soret band. Oxygenated hemoglobin has additional peaks of absorption in the visible range at 540 nm and 575 nm (Figure 3-4). When all oxygen is removed from the medium, these peaks disappear and are replaced by a less sharp maximum at 555 nm. Thus arterial blood is bright red; venous blood, containing less oxygen, appears more blue.

Carbon dioxide transport

Carbon dioxide is transported in bloods as bicarbonate, the anion paired with Na$^+$, and does not require a transport pigment. The equilibrium between gaseous CO_2 and carbonic acid, a necessary step in bicarbonate formation is catalyzed by the enzyme carbonic anhydrase:

TABLE 3-1. Molecular weights of hemoglobins and myoglobins.

	Hemoglobin	Myoglobin
holothurians		
(Sclerodactyla) briareus	23,600	–
Molpadia arenicola	35,000	–
polychaetes		
Glycera dibranchiata	18,200	–
Abarenicola pacifica	2,600,000	16,000
bivalve mollusks		
Phacoides pectinatus	–	14,580
Arca pexata	33,600	–
Anadara inflata	72,000	–
Cardita floridana	3,000,000	–
chiton		
Acanthopleura	–	16,000
granulata	–	34,600
gastropods		
Aplysia depilans	–	22,580
Busycon canaliculatum	–	31,000

From Prosser, 1973; Garlick & Terwilliger, 1977.

$$CO_2 + H_2O \rightleftharpoons H_2CO_3 \rightleftharpoons H^+ + HCO_3^-$$

This enzyme is abundant in gills of many marine invertebrates. Carbon dioxide also influences the transport of O_2 by contributing to the acidity of blood. As CO_2 pressure increases, [H+] increases, and hemoglobin becomes less capable of binding O_2. Thus the release of O_2 is favored in the tissues, and uptake is favored in the gills where CO_2 concentration is lower, an effect promoting efficiency of O_2 transport. This effect is shown by the displacement of the oxygen-dissociation curve as in Figure 3-2.

Myoglobin

Myoglobin is a muscle pigment very similar to hemoglobin but with smaller molecular weight and only one heme group per molecule. It has similar oxygen-binding properties and has been studied extensively as a model for hemoglobin. In the polychaete

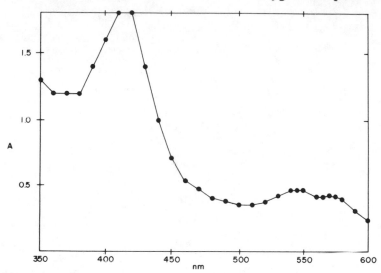

Figure 3-4. Absorption spectrum of hemoglobin from a polychaete worm, *Glycera dibranchiata,* often sold in bait stores as the bloodworm. Results of a class exercise.

Arenicola, hemoglobin (Figure 3-5) releases O_2 to the myoglobin of body wall muscles. In the chiton, *Cryptochiton stelleri,* the myoglobin has a greater affinity for O_2 than the hemocyanin, which suggests that it transfers O_2 from blood to muscle (5). Myoglobin extracted from the radula muscle of another chiton, *Acanthopleura granulata,* was resolved into two substances, the minor component with twice the molecular weight of the major component (9). The range of molecular weights is narrow (Table 3-1), with the smallest myoglobin, from *Phacoides,* having 136 amino acid residues, and one of the larger, from *Aplysia,* having 205 residues. The sizes of hemoglobins of two other bivalves, *Arca* and *Anadara,* suggest multiples of a myoglobin slightly larger than that of *Phacoides.*

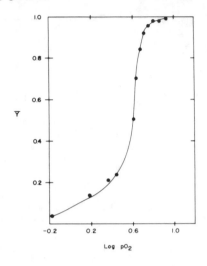

Figure 3-5. Oxygen equilibrium curve of lugworm *Arenicola* hemoglobin, in phosphate buffer at pH 7.6, temperature 22°. (From Waxman, 1971).

Chlorocruorin

Another compound resembling hemoglobin even more closely is the blood pigment of serpulids and sabellids (Polychaeta) called chlorocruorin. It has a formyl group replacing a vinyl group on the side chain of the heme. This is sufficient to make chlorocruorin appear a darker hue of red. In several species of serpulids, both chlorocruorin and hemoglobin are present. Others of this group have only one or the other. Chlorocruorin from *Serpula vermicularia* has a molecular weight of 3,000,000. Properties are similar to hemoglobins, except that higher O_2 concentrations are required for saturation.

Hemerythrin

Hemerythrin is the O_2 transport and storage pigment of sipunculids, priapulids, and brachiopods. It is a protein containing iron, but lacking heme. Hemerythrin occurs in special cells called coelomocytes, and is pink when oxygenated and colorless when deoxygenated. Molecular weights range from 66,000 in *Sipunculus* to 107,000 in *Golfingia*. Iron is present as Fe^{++} and Fe^{+++} in equilibrium, each pair of atoms holding one molecule of oxygen (5). In hemerythrin of *Golfingia gouldii* there are 16 g-atoms of Fe per mole of protein. On reduction of disulfide bonds, the protein dissociates into units of 13,500 molecular weight, which suggests that it is an octomer with two atoms of Fe in each monomer. Hemerythrin of the brachiopod *Lingula reevi* displays strong O_2 affinity and a strong Bohr effect, comparable to hemoglobin of an active animal. Brachiopods, however, are sedentary animals with low rates of O_2 consumption. Blood pigments of relatively inactive animals are presumed to function as oxygen-storage compounds. The supply of stored O_2 is often not great enough to see these animals through normal periods of O_2 lack. It merely retards the onset of an inevitable shift to anaerobic metabolism. For example, the hemoglobin of *Arenicola* functions to provide 15 to 45 percent of the oxygen consumed while the animal is ventilating its burrow, but it holds only enough oxygen to last twenty minutes when the water becomes hypoxic at low tide (4).

This account of the major and minor oxygen-carrying pigments has shown that they function in a similar manner in a great variety of animals. Details of their molecular architecture are accumulating to the point where some pigments are now being used as models for intensive study of the transport process, and some as data to support ideas on biochemical evolution.

Mya arenaria

CHAPTER 4　Digestion

The food of marine invertebrates consists largely of plankton, fragments of plankton with associated microorganisms, and dissolved organic substances. Plankton is a collective term for a great variety of plants and animals, mostly very small and with limited powers of self-locomotion. A few animals feed on larger algae, and some prey on other invertebrates, such as the sea stars and drills that attack bivalves.

Ingestion and digeston

Mechanisms of ingestion are extremely varied. They include suspension-feeding with the aid of cilia, setae, and sheets of mucus, and deposit-feeding, as well as the more obvious methods of larger predators (7). What all these food habits and feeding activities have in common is that they are preludes to digestion, a chemical breakdown of food required to make it useful. Digestion is the work of hydrolytic enzymes.

Hydrolysis is the splitting of large food molecules into smaller ones that can be absorbed and metabolized further. Water is required not only as a solvent, but also as a reagent. The atoms of water are incorporated into the product molecules. Hydrolytic enzymes, like all enzymes, are protein catalysts constructed by living cells.

Sponges, flatworms, coelenterates, and lamellibranch mollusks depend heavily on phagocytosis, the ingestion of particles by cells, followed by intracellular digestion (6). As animals increase in complexity of body plan, more secretion of enzymes into the lumen of a digestive tract occurs.

Enzyme nomenclature

Improved communication between scientists interested in digestive enzymes has resulted from use of a system of enzyme nomenclature proposed by the "Enzyme Commission" of the International Union of Biochemistry (3). In this system, each enzyme has a four-part "EC" number, a systematic name, and a recommended trivial name. For example, all hydrolases are EC 3, glycosyl hydrolases are EC 3.2, glycoside hydrolases EC 3.2.1, and the starch-digesting enzyme amylase is designated EC 3.2.1.1, with systematic name 1, 4-a-D-glucan glucanohydrolase. Definite rules apply to the naming and classification of enzymes, so that new ones can be incorporated into the scheme as they are discovered. In general, the names are assigned on the basis of the substrate upon which the enzyme is most active, the type of reaction catalyzed, and the direction of reaction that has been experimentally demonstrated.

Principal digestive enzymes

Among the principal digestive enzymes detected in marine invertebrates are:

lipase = EC 3.1.1.3 = triacylglycerol acyl-hydrolase
a-amylase = EC 3.2.1.1 = 1, 4-a-D-glucan glucanohydrolase
trypsin = EC 3.4.21.4 (a serine proteinase)
pepsin = EC 3.4.23.1 (an acid proteinase)

Extracts of the gut of most invertebrates display proteinase, amylase, and lipase activity.

The amylases act on starch, glycogen, and related poly-saccharides. These are high-molecular-weight compounds consisting of 300 or more glucose units. The a-amylases cleave the macromolecule at random in the middle, initially producing oligosaccharides of 6 to 7 glucose units, skipping the 1,6 bonds, and ultimately producing maltose fragments. Other names for a-amylase are diastase, ptyalin, and glycogenase. The β-amylases do not concern us here since they occur predominantly in plants. However, their action is to remove successive maltose units from the nonreducing ends of the chains: β-amylase = EC 3.2.1.2 = 1,4-a-D-glucan maltohydrolase.

Detection of hydrolysis

After choosing the enzyme to be studied, the next step is selecting a good assay method. Starches react strongly with molecular iodine to form a deep blue complex. When a-amylase attacks a starch, there is a rapid decrease in the starch-iodine reaction as the macromolecule is fragmented. The quantity of "reducing sugar" increases at the same time, as fragments with free aldehyde groups appear (Figure 4-1). These reducing sugars are oxidized by aldehyde reagents such as Benedict's solution to yield a red precipitate of cuprous oxide:

$$2\,Cu(OH)_2 + RCHO \longrightarrow Cu_2O\downarrow + RCOOH + 2\,H_2O$$

Thus the progress of starch digestion may be followed by decrease in color of the iodine complex or increase in color of cuprous oxide. With a spectrophotometer, precise measurement of color change is possible; amylase activity has been detected in many species by these methods.

Substrates like starch, with high and variable molecular weights, may be degraded in such a variable manner that the rate of formation of a well-defined product is the best way to express the rate of enzymatic breakdown. The abilities of

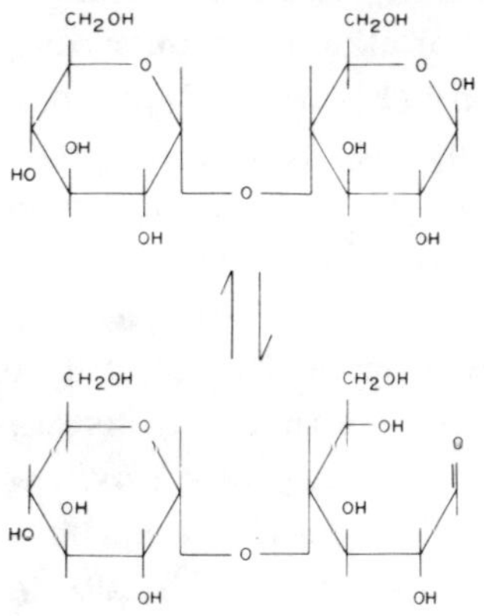

Figure 4-1. Structures of maltose in a closed ring form and in a free, reactive aldehyde form, the latter responsible for "reducing sugar" properties.

twenty-two species of marine invertebrates to digest twenty-nine carbohydrates were reported as fractions of the total reducing sugars possible from complete hydrolysis (4). Glycogen was digested from 10 percent to 68 percent in 15 minutes by enzymes in the extracts of the various species. Amylose and laminaran were digested by most species, and chitin by a few, but twenty-five other carbohydrates were resistant to hydrolysis. An investigator who had found activity and established specificity of a digestive enzyme would properly extend his study to the kinetics and chemistry of the enzyme, but prior to 1965 very few had done so (1).

Calculation of specific activity

In some gastropods and most lamellibranch mollusks, amylase is found in a gelatinous rod called the "crystalline style" that is formed by a style sac. It is dissolved at the free end, liberating the enzyme into the stomach.

Activity of a-amylase in the style of a bivalve can be

compared with purified a-amylase from other sources, by means of the starch-iodine or alkaline copper reactions.

In one experiment (Figure 4-2), starch hydrolysis was catalyzed by pure a-amylase from a bacterium, and in another it was accomplished by an extract of the style of the clam *Mya arenaria.* Curve A shows the increase in light absorbance by Cu_2O as the pure amylase produced reducing sugar, and curve B shows the decrease in absorbance by starch-iodine complex as style amylase did its work. Both reactions slowed down after 1-2 minutes. Reactions ordinarily slow down as substrate is used up, so rates are usually based on the earliest measurable changes. The rates in these examples are $0.100\Delta A/min$ for reaction A, and $0.200\Delta A/min$ for reaction B.

In order to compare these rates properly, they must be expressed as numbers of moles of maltose produced in unit time by unit weight of enzyme. First, the ratio between absorbance and change in concentration must be determined. In experiment B, the absorbance of the starch-iodine sample before hydrolysis began (zero-time sample) was 0.540. The concentration of starch in this sample was $0.0877\,g/l$.

Concentrations are better expressed on a molar basis than on the basis of g/l. The molecular weight of maltose is 342.3, and each gram of starch produces finally a little more than 1 g of maltose, because of the water added in hydrolysis. The molecular weight of water is 18.0, and in starch digestion each mole of maltose accepts one mole of water, which makes the molecular weight of maltose, as it exists in starch, 324.3. Thus the concentration of maltose expected from $0.0877\,g/l$ starch is $2.70 \times 10^{-4}M$. The ratio between absorbance and molar concentration is called the molar absorptivity, symbolized by epsilon (ϵ). For most colored substances, ϵ is a large number, about 10^3 to 10^4, and it is constant at a specific wavelength, generally chosen at a maximum of light absorption. In this case of the starch-iodine reaction:

$$\epsilon = 0.540/2.70 \times 10^{-4}\ M = 2.00 \times 10^3\ A/M$$

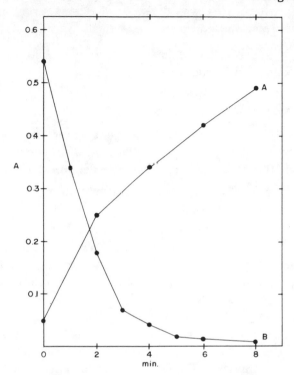

Figure 4–2. Amylase activity of style of *Mya arenaria* and of purified enzyme. Curve *A:* absorbance at 420 nm of 1-ml samples of reaction mixture plus 5 ml Fehling's solution; mixture contained 25 ml starch 20 mg/ml and 0.5 ml pure *a*-amylase 20 mg/ml. Curve *B:* absorbance at 620 nm of 0.2-ml samples of reaction mixture plus 0.2 ml 2N HCl, 1.0 ml I_2-KI solution, 10 ml water; mixture contained 5 ml starch 10 mg/ml, 4 ml 0.1*M* phosphate buffer pH 6.0, and 1 ml style extract 50 mg/ml.

The reaction rate of the *Mya* amylase can then be expressed as:

$$\text{Rate} = \frac{0.200 \Delta A/\text{min}}{2.00 \times 10^3 \ \Delta A/\Delta M} = 1.00 \times 10^{-4} \Delta M/\text{min}$$

In order to convert this rate of concentration change into a rate of production of maltose, the volume of the reaction mixture is included in the calculation:

Rate = 1.00×10^{-4} mole/l-min $\times 10^{-2}$l =
1.00×10^{-6} mole/min

The weight-specific rate takes into account that the reaction mixture contained 50 mg of style material:

Rate = 1.00×10^{-6} mole/min per 50 mg =
2.00×10^{-5} mole/min per g

The style of *Mya arenaria* is 80 percent water, and one-half of the dry material is protein (2). Therefore each gram of style represents 100 mg protein, and the rate becomes:

Rate = 2.00×10^{-7} mole/min – mg protein =
0.20μ mole/min – mg protein

This rate is very close to the ideal way of expressing specific activity of enzymes as given by the Enzyme Commission (3). Specific activity is the number of units per mg protein in an enzyme preparation, and the unit is "that amount which will catalyze the transformation of one micromole of the substrate per minute under standard conditions." For digestive enzymes, the number of micromoles of product is generally more attainable than the micromoles of substrate. The specific activity of the bacterial amylase (Figure 4–2) by a similar sequence of calculations is:

Rate = 3.32μ mole/min – mg amylase

This suggests that the specific activity of the molluscan-style amylase is about 6 percent of the specific activity of the microbial amylase. This is a minimum estimate, because the style is known to contain other enzymes besides amylase.

The advantage of varying substrate concentration

In order to compare the effectiveness of the same digestive enzyme from different species, a common procedure is to measure rate of hydrolysis at a single, high concentration of substrate. The enzyme is presumed "saturated" or acting at near its maximum capacity. This approach is defective for two reasons: (1) if only one concentration of substrate is used, it remains unknown whether a higher concentration might produce a still higher rate; and (2) the concentration chosen for the assay may be much above or below any that the enzyme ever encounters *in vivo*, and therefore the result is merely an artifact.

Maximal velocity and the Michaelis constant

These defects are remedied and information is greatly augmented when activity is measured at several substrate concentrations over a suitable range. The data can be used to calculate a theoretical maximum velocity (V), and a substrate concentration at which the rate is exactly one-half maximal, called the Michaelis constant (Km). With these two points known, a rate of catalysis may be estimated for any desired concentration, including those that normally occur in the animal.

A good example of these concepts in action is a study on lipase (EC 3.1.1.3) of the surf clam *Spisula solidissima* (8). Rates of hydrolysis of triolein ($C_{57}H_{104}O_6$) by style extract were determined at seven substrate concentrations. Results were given in nanoequivalents of oleic acid released per minute per mg protein in the enzyme preparation.

Lipase of *Spisula* was prepared by simply freezing and drying the crystalline styles, then homogenizing the dry material in buffer (3.33 g/l), and centrifuging briefly to remove undissolved material. The greatest activity was found at pH 8.0 and 20° C. *Spisula* lipase was equally active on triolein, methyl oleate, and methyl elaidate. At an intermediate substrate concentration the style enzyme displayed activity equivalent to 63 percent that of a lipase from the skate *Raja erinacea*, but only 2 percent that of hog pancreas enzyme. Rates (v) increased with substrate concentration (S) in an hyperbolic manner (Figure 4-3).

This relation suggests that the rate at any moment is

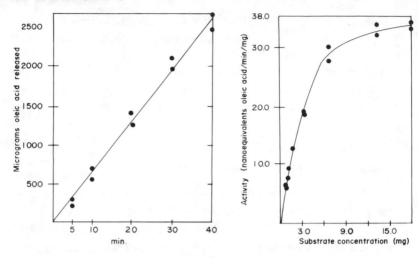

Figure 4-3. Velocity of hydrolysis of triolein, catalyzed by lipase from crystalline style of clam *Spisula*, increases with substrate concentration and appears to approach a maximum.

proportional to the substrate concentration, up to a point where the amount of enzyme present becomes limiting. These are precisely the conditions that allow calculation of a Michaelis constant (Km) and maximal velocity (V). A graphical method of estimating these values, which are fundamental properties of an enzyme, is shown in Figure 4-4 (5). A plot of S/v vs. S gives $Km = 0.40$mM and $V = 0.044\mu$mole/min per mg protein in the *Spisula* style.

Future work

Determination of properties such as pH and temperature optimum, substrate specificity, Km, and V allow one to make meaningful comparisons of the same enzyme from different species. There are still insufficient data on digestive enzymes from marine invertebrates, and we must await new work that goes well beyond mere evidence that an enzyme is present. However, the work requires good direction, not massive labor.

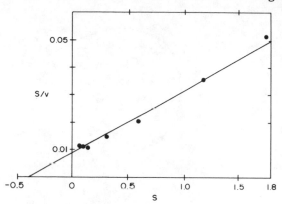

Figure 4-4. Relation of the ratio *S/v* to substrate concentration (*S*) for reaction catalzyed by style lipase in Figure 4-3. Straight line extrapolates to *-Km* = -0.40 m*M* triolein, and *V* is estimated at 0.044 μmole/min per mg enzyme from the slope of the line, which is 1/*V*.

For example, extensive purification is often unnecessary for determining the properties listed above. Specific activity in the conventional sense, based on mg of protein, rises with purification, and can be merely a measure of success in obtaining a pure enzyme. Maximal activity (*V*) per unit weight of tissue is easier to obtain and has the virtue of referring more directly to the living animal.

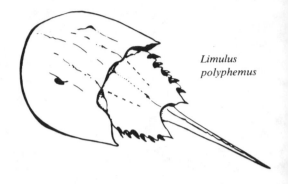

Limulus polyphemus

CHAPTER 5 Intermediary metabolism

A barnacle sweeps the immediate water mass to extract food particles; it less obviously removes oxygen from the water and releases carbon dioxide. These are the gross manifestations of a multitude of chemical reactions occurring continuously within the animal's body. These interrelated reactions make up intermediary metabolism. The controlled release of the chemical bond energy of foods is called catabolism, while the sum of all biosynthetic processes is called anabolism.

Glycolysis

The most important and best-known aspect of catabolism is the system of carbohydrate degradation known as glycolysis. According to the standard scheme, derived from studies on yeast and vertebrate muscle, glycolysis begins with starch or glycogen and proceeds through ten phosphorylated intermediates to pyruvic acid. Later steps are characteristic of species and influenced by oxygen supply. Each reaction in the sequence is catalyzed by a specific enzyme, and the complete set occurs in every animal. For example, gill tissue of the shore crab *Carcinus maenas* proved to have measureable activities of fourteen enzymes of glycolysis and the pentose phosphate pathway (16). This type of result is the general rule, rather than the exception. Occasional reports of the absence of an enzyme of glycolysis should be taken as evidence of inadequacy of the methods, not of the animal.

Enolase

Enolase is an enzyme of glycolysis which has received more than passing attention. The systematic name is 2-phospho-D-glycerate hydrolyase (EC 4.2.1.11), and the reaction is:

$$
\begin{array}{ccc}
\text{COOH} & & \text{COOH} \\
| & -H_2O & | \\
\text{HC-O-PO}_3\text{H}_2 & \rightleftharpoons & \text{C-O-PO}_3\text{H}_2 \\
| & & \| \\
\text{CH}_2\text{OH} & \text{enolase} & \text{CH}_2
\end{array}
$$

2-phospho-D-glyceric acid   phosphoenolpyruvic acid (PEP)

An unusually thorough study concerned the enolases of six species of mollusks, four gastropods and two pelecypods (11). The results of this work (Table 5-1) indicated a relatively

TABLE 5-1. Properties of enolases from mollusks. Specific activity in μmoles/min per mg protein. Michaelis constant in mM.

	Molecular Weight	Specific Activity	Km (PGA)
GASTROPODA			
Busycon canaliculatum	79,000	40.0	0.0345
Thais lapillus	80,000	11.0	0.0230
Littorina littorea	95,000	9.3	0.0202
Ilyanassa obsoleta	84,000	9.7	–
PELECYPODA			
Mytilus edulis	84,000	4.7	0.0580
Mercenaria mercenaria	80,000	8.1	0.0530

(From Hanlon *et al.*, 1971)

narrow range of molecular weights and Michaelis constants. Specific activities also varied little, except for the more active enolase of the large whelk *Busycon*. All of these enzymes were inhibited by *p*-mercuribenzoate, and all except *Busycon* enzyme were protected from this inhibition by presenting them with substrate, phosphoglyceric acid, before adding the inhibitor. This suggests the importance of sulfhydryl groups for a functional active site. Cysteine and dithiothreitol were added to the extract during the purification process to prevent oxidation of sulfhydryl groups on the enzymes. Although the enolases from these six species are quite similar, it is clear that they are not identical proteins. Substances that play the same role in metabolism and show signs of chemical kinship have been designated "isologues" (6). This distinguishes them from homologues, which are identical molecules, and analogues, chemically unrelated substances that play the same physiological role. The ATP's from different species are homologues, while the luciferins from different species are analogues. The degree of difference between isologues suggests the amount of divergence of species in evolution. In the case of the enolases, the bivalves have lower specific activity and higher Michaelis constants than the gastropods, a distinction that complements their morphological differences.

The pentose phosphate pathway

A variable fraction of the glucose used in glycolysis proceeds *via* an alternate series of reactions known as the pentose phosphate pathway. In this pathway, an oxidative step occurs earlier than in standard glycolysis, and the coenzyme required is NADP rather than NAD. The crab *Carcinus maenas* has activities of several of the enzymes of this alternate pathway and has more NADP than NAD in its gill tissue (16), suggesting dependence on this pathway.

The citric acid cycle

Having reached the stage of pyruvic acid, carbohydrate degradation may take several directions. Under aerobic conditions, the most probable is decarboxylation to form acetylcoenzyme A, which then condenses with oxaloacetate to form citric acid. Through the ten steps of the citric acid cycle, hydrogen atoms are removed and two more decarboxylations occur, completing the dismantling of pyruvate. The H is ultimately oxidized to water by means of the cytochrome system. The overall process is:

$$CH_3CO\ COOH + 2\frac{1}{2}O_2 \rightarrow 3CO_2 + 2H_2O$$

Intermediates and enzymes of the citric acid cycle, originally proposed by Hans A. Krebs in 1937, have been demonstrated in many species of marine invertebrates. Often, however, there have been suggestions that the cycle is incomplete, only partially functional, or not of major importance. For example, oxygen consumption of the mantle tissue of the oyster, *Crassostrea virginica*, was stimulated very little (16 to 20 percent) by cycle intermediates, succinate dehydrogenase activity was barely measureable, and aconitase activity was absent (12). Several factors can help to explain failures to detect certain dehydrogenases that are universally present.

The cellular environment of enzymes in marine invertebrates includes a total osmotic concentration of about $1.10\,M$. Therefore, it is not surprising to find that a dehydrogenase fails to work well in ordinary $0.050\,M$–$0.067\,M$ phosphate buffer. Mitochondrial enzymes are essential to the general process of oxidative phosphorylation. Mediums of at least $0.30\,M$ were required to obtain good rates of phosphorylation from isolated mitochondria of the blue crab, *Callinectes sapidus* (2).

The "standard" assay of an enzyme may also employ a buffer with pH too far from optimum. Oxidation of lactate by lactate dehydrogenase, commonly measured at pH 9.0, was catalyzed best at pH 7.5 by enzyme from a barnacle, *Balanus nubilus;* the barnacle dehydrogenase was rapidly inactivated at more alkaline pH (5). Temperatures above 20° C may produce heat denaturation of enzymes from marine invertebrates. Sometimes substrate concentrations much higher than expected must be used. A brachiopod dehydrogenase displayed no lactate oxidation with $0.040M$-$0.100M$ lactate, but activity was readily measured at $0.300M$-$0.500M$ (9). In summary, inappropriate choice of pH, osmotic strength, temperature, or substrate concentration can all result in lack of detectible activity, so it is wise to use a range of each factor before reporting a "missing" enzyme.

Lactate dehydrogenase

The phenomenon of oxygen debt, or increased O_2 consumption after an anaerobic period, has been observed in several marine invertebrates. The cellular basis is the oxidation of accumulated lactic acid, and the extra energy required to resynthesize glycogen from lactic acid. The reaction is catalyzed by lactate dehydrogenase (EC 1.1.1.27):

$$CH_3 \, CO \, COOH + NADH + H^+ \rightleftarrows CH_3 \, HCOH \, COOH + NAD^+$$
$$\text{pyruvic acid} \qquad\qquad\qquad \text{lactic acid}$$

Formation of lactic acid is greatly favored by the equilibrium, but lactic acid can be oxidized to pyruvic acid if pyruvate and reduced NAD concentrations are very low. The progress of the reaction depends in part on the enzyme's "readiness to combine." This is expressed in simple form by Michaelis constants, some of which are given in Table 5-2. The values of Km for pyruvate range from $0.07mM$ in the horseshoe crab $Limulus$ to $8.9mM$ in a brachiopod, $Terebratulina$, while the values of Km for lactate extend over a higher range, $2.5mM$ to $400mM$, in lobster and brachiopod, respectively. According to these results, $Limulus$ is the most likely to produce lactic acid in anaerobiosis, and the lobster $Homarus$ is best equipped to oxidize it. The brachiopods have the highest values of both constants, which

TABLE 5-2. Michaelis constants of lactate dehydrogenases from some marine invertebrates (mM).

	Pyruvate	Lactate	
MOLLUSCA			
Mercenaria mercenaria	0.61	250	(D)
Crassostrea virginica	0.54	120	(D)
Cardium edule	0.16	16	(D)
		48	(L)
Loligo pealei	0.64	–	
Haliotis cracherodii	0.16	1.9	(D)
CRUSTACEA			
Homarus americanus	0.40	2.5	
Balanus nubilis	1.5	47	(D)
MEROSTOMATA			
Limulus polyphemus	0.07	3.8	(D)
ANNELIDA			
Nereis virens	0.13	3.1	(D)
ECHINODERMATA			
Sclerodactyla briareus	2.2	35.1	
BRACHIOPODA			
Terebratulina septentrionalis	8.9	290	
Glottidia pyramidata	2.6	400	

(From Ellington and Long, 1978, and various other sources.)

suggests that neither formation nor oxidation of lactate proceeds rapidly in their tissues.

The lactate dehydrogenases of marine invertebrates differ in substrate specificity. Some species contain an enzyme that is active only on L-lactate, and others an enzyme that produces and oxidizes only the stereoisomer D-lactate. Stereoisomers are compounds with atoms grouped in an asymmetric arrangement such that their images are not superposable. Prefixes D and L refer to the direction of optical rotation in the reference compounds, the stereoisomers of glyceraldehyde, but not necessarily to the optical rotation of derived compounds. The stereoisomers of glyceraldehyde and of lactic acid are shown in Figure 5-1. Animals with specificity for L-lactate include members of at

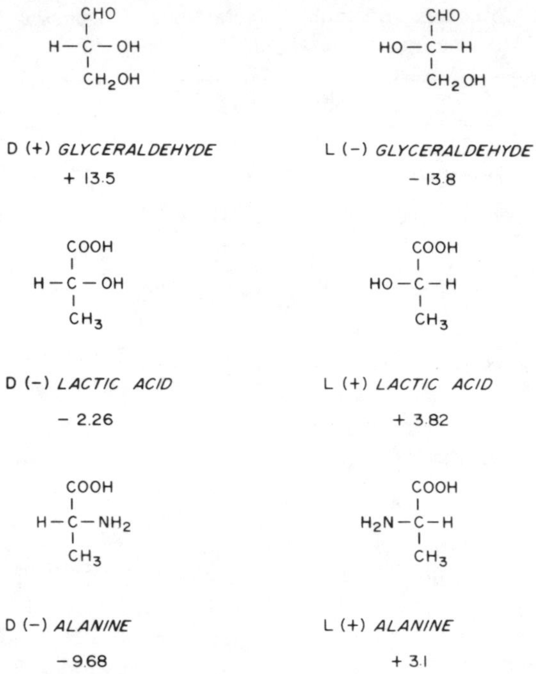

$$CHO$$
$$H-C-OH$$
$$CH_2OH$$

D (+) *GLYCERALDEHYDE*
+ 13.5

$$CHO$$
$$HO-C-H$$
$$CH_2OH$$

L (−) *GLYCERALDEHYDE*
− 13.8

$$COOH$$
$$H-C-OH$$
$$CH_3$$

D (−) *LACTIC ACID*
− 2.26

$$COOH$$
$$HO-C-H$$
$$CH_3$$

L (+) *LACTIC ACID*
+ 3.82

$$COOH$$
$$H-C-NH_2$$
$$CH_3$$

D (−) *ALANINE*
− 9.68

$$COOH$$
$$H_2N-C-H$$
$$CH_3$$

L (+) *ALANINE*
+ 3.1

Figure 5-1. Structural formulas showing configuration of D-glyceraldehyde and L-glyceraldehyde, the stereoisomers that serve as reference compounds in assigning D and L to other pairs. Numbers indicate degree and direction of optical rotation. Note that direction is independent of configuration in derived compounds.

least seven invertebrate phyla: Porifera, Cnidaria, Nematoda, Brachiopoda, Annelida, Arthropoda, and Echinodermata. Animals with a dehydrogenase specific for D-lactate include all Mollusca, polychaete annelids, *Limulus*, barnacles, and all Arachnida (14). The distribution of these dehydrogenases with opposite steric requirements suggests that one or the other enzyme became fixed by heredity in each line of descent. Lactic acid is either excreted or converted back to pyruvic acid by the same enzyme, so the isomeric form of lactic acid does not affect other reactions or pathways.

Succinate production in anoxia

In some species where enzyme properties indicate very low rates of lactic acid production, an alternate end product of anaerobiosis, succinic acid, is formed. Succinate formation occurs by means of carbon dioxide fixation with pyruvate or phosphoenolpyruvate to form a four-carbon acid, either malate or oxaloacetate. Then by reversal of the citric acid cycle through two or three steps, fumarate is converted to succinate, which accumulates. Presumably, after a period of oxygen lack, accumulated succinate can be oxidized through the citric acid cycle when oxygen is again available. For example, succinate in tissue of *Mytilus edulis* after forty-eight hours out of water was twelve times greater than that of controls, 2.94μmoles/g and 0.24μmoles/g, respectively (18). When most marine invertebrates are taken from the water, their gaseous exchange comes to a halt. The longer they are out of the water, the more anaerobic they become, and the more succinate is produced. Even after death of the animal, the enzymes of the succinate pathway may continue to function for hours. For animals such as bivalves that sometimes endure long periods of anoxia, succinate formation appears preferable to lactate formation. There is extra ATP production in the conversion of fumarate to succinate, and there may be less change in cellular pH because succinic is a weaker acid than lactic.

Alanine from glycolysis

The citric acid cycle serves as a crossroads of carbohydrate, fat, and protein metabolism. Carboxylations in animal cells require chemical bond energy in the form of high-energy phosphate compounds or reduced coenzymes. The supply of PEP and reduced NADP from glycolysis allows pyruvate to be carboxylated to form oxaloacetic and malic acids; pyruvate may also be converted to alanine. Transaminations involving pyruvate, a-ketoglutarate, and oxaloacetate have been measured in several mollusks (8). The amino acids related to these carbohydrate derivatives are alanine, glutamic acid, and aspartic acid. Conversion of alanine to pyruvate in the small bivalve *Solemya velum*, for example, was very rapid, indicating exceptionally high activity of L-alanine: 2-oxoglutarate aminotransferase (EC 2.6.1.2):

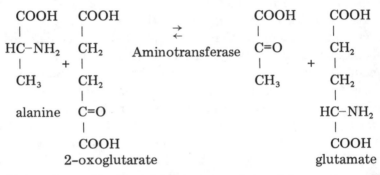

If alanine from the diet should accumulate, this enzyme permits its conversion to pyruvate. The carbon of pyruvate may take several pathways (Figure 5-2), including three leading to formation of intermediates of the citric acid cycle. The hydrogen and electrons from the citric acid cycle enter the cytochrome system. The first report of cytochromes by Keilin in 1925 established these intracellular pigments as ubiquitous. Inhibition of oxygen consumption by 0.2–10.0 mM cyanide implies a cytochrome oxidase in many species where it has not been demonstrated directly. Cytochrome c was isolated and purified from an oyster, a squid, and a prawn, and each was isologous according to absorption spectrum and reactivity with cytochrome oxidases from various sources (17).

Amino acid oxidation

Part of the oxygen consumption of marine invertebrates may be a manifestation of amino acid oxidation. Many species of mollusks have an L-amino acid oxidase active on arginine,

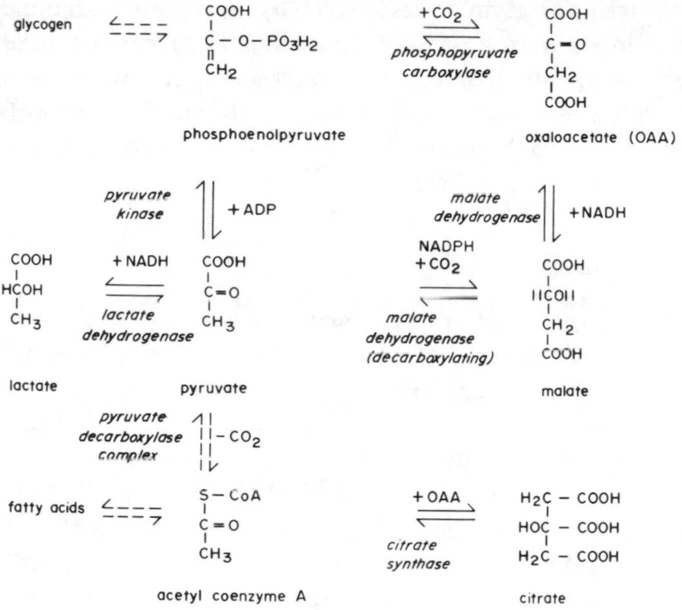

glycogen ⤙- - -⇀

COOH
|
C − O − PO$_3$H$_2$
‖
CH$_2$

phosphoenolpyruvate

+CO$_2$ ⇌
phosphopyruvate
carboxylase

COOH
|
C − O
|
CH$_2$
|
COOH

oxaloacetate (OAA)

pyruvate
kinase   + ADP

malate
dehydrogenase   +NADH

NADPH
+CO$_2$

COOH
|
HCOH
|
CH$_3$

lactate

+ NADH ⇌
lactate
dehydrogenase

COOH
|
C=O
|
CH$_3$

pyruvate

malate
dehydrogenase
(decarboxylating)

COOH
|
HCOH
|
CH$_2$
|
COOH

malate

pyruvate
decarboxylase
complex   − CO$_2$

fatty acids ⤙- - -⇀

S − CoA
|
C = O
|
CH$_3$

acetyl coenzyme A

+ OAA ⇌
citrate
synthase

H$_2$C − COOH
|
HOC − COOH
|
H$_2$C − COOH

citrate

Figure 5-2. Summary of possible fates of pyruvate derived from glycolysis. Pyruvate may be carboxylated to form a 4-carbon acid, as may also happen to its immediate precursor PEP. Or it may be reduced to lactate, or it may be decarboxylated to form acetyl CoA which participates in citrate synthesis.

ornithine, lysine, and histidine (7). Reactions of this type produce hydrogen peroxide:

R–CH COOH
|
NH$_2$ $\xrightarrow{O_2}$

amino acid

R–C COOH
‖
NH

imino acid + H$_2$O$_2$

$\xrightarrow{H_2O}$

R–C COOH
‖
O

keto acid +
NH$_3$

Catalase (EC 1.11.1.6) accelerates breakdown of the peroxide:

$$2H_2O_2 \longrightarrow O_2 + 2H_2O$$

When higher rates of O$_2$ consumption are associated with higher catalase activity, as in *Crassostrea virginica*, the probability of amino acid oxidation is strengthened (1). Animal proteins are constructed exclusively of L-amino acids, but the optical isomers are of widespread occurrence in the free form (3). It is probable that the D-amino acids in animals are derived from plants and microorganisms in the diet.

Protein synthesis

Anabolic processes are less well-known. Growth continues throughout life in many marine invertebrates, although it is much slower in older animals. Even during apparent adult stability, mucus, enzymes, and hormones are continuously synthesized. Most of these substances are not yet completely characterized. In developing sea urchins, protein synthesis is accomplished by the standard apparatus of ribosomes and three major types of RNA, in accordance with the biochemical unity principle (1). Adult lobsters synthesize new soluble proteins continuously at a low rate that can be measured as the rate of incorporation of ^{14}C-labelled amino acids (4). An unusual feature of protein synthesis in sea anemones is the presence of 2-aminoethylphosphonic acid (AEP), up to 1 percent of total protein. This compound contains a carbon-phosphorus bond, unknown in natural substances until found in *Anthopleura elegantissima* (13).

$$
\begin{array}{c}
H_2C\!-\!NH_2 \\
| \\
CH_2 \\
| \\
O\!=\!P\!-\!OH \\
| \\
OH
\end{array}
$$

2-aminoethyl phosphonic acid

AEP occurs in phospholipids, glycerol ester, and free form, as well as in protein. Compounds with C-P bonds were found in some species of six phyla: Porifera, Coelenterata, Mollusca, Annelida, Arthropoda, and Echinodermata; but only the anemones had large amounts in proteins (15). Many more special aspects of intermediary metabolism may be expected to come to light when more species of marine invertebrates are studied.

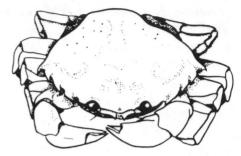

Carcinus maenas

CHAPTER 6 **Total metabolism**

Any dense aggregation of animals, such as a scattering layer of zooplankton or an oyster bed, must be a little warmer than its surroundings, because of metabolic heat production. The "fire of life" (5) keeps animal body temperatures always a little greater than their immediate environment. Metabolism consists of the controlled release of energy from foods by hundreds of enzyme-catalyzed reactions and the use of some of that energy to build new animal-substance.

Heat production

No energy transformation, according to the second law of thermodynamics, is perfectly efficient. Some fraction of the bond energy moved about in a chemical reaction always appears as heat. When an animal is in a steady state, the rate of energy capture in ATP is equal to the rate of ATP hydrolysis, as it is used to drive muscles, cilia, ion pumps, and synthetic reactions. ATP hydrolysis is the source of most of the heat that appears in metabolism. These processes all occur "behind the scenes," inside of living cells and tissues, but they are manifested in whole-animal phenomena such as gaseous exchange, nitrogen excretion, and heat production.

Calorimetry has been used to determine metabolic rates of animals since Lavoisier and Laplace (7) found in 1780 that respiration of guinea pigs is nearly equal to the combustion of carbon in both heat production and gaseous exchange. A guinea pig used by Lavoisier and Laplace endured ten hours at 0° C in an ice calorimeter. Heat was calculated from the weight of water from melted ice. Another animal was immersed in mercury on the way into and out of a bell-jar respirometer, and was removed after 1¼ hours when it began breathing with difficulty in the atmosphere of decreasing oxygen and increasing carbon dioxide. The experiments gave an average heat production of 3,218 cal/hr, and an oxygen consumption of 691.5 ml/hr, corrected to 0° C. Thus their guinea pig had a heat equivalent of oxygen of 4.65 cal/ml. Combustion of food substances in a bomb calorimeter gives ratios of 4.5 to 6.0 calories per ml oxygen.

Indirect calorimetry

The confirmation of this constant relation between heat and oxygen led to the generalization that animals obey the laws of thermodynamics, just as inanimate systems do. It also justified indirect calorimetry, the measurement of total metabolism by means of gaseous exchange alone. Indirect calorimetry consists of measuring oxygen uptake, assuming a constant heat equivalent such as 4.65 cal/ml, and then calculating rates of heat production. When animals obtain all of their energy needs by using oxygen for oxidation of food substances the methods are equally valid. Then greater ease of construction and operation makes respirometers preferable to calorimeters. All animals are aerobes, in the sense that even among parasites and sediment inhabitants, no species has been found to complete its life cycle in the total absence of oxygen. Therefore, the assumption is largely correct.

Agreement between the two measures was not really tested in animals other than mammals until recently. The first invertebrates to be placed in a calorimeter were earthworms used by A. V. Hill (4), and fertilized eggs of a sea urchin used by O. Meyerhof (8). Hill constructed a "differential microcalorimeter", which consisted of a pair of vacuum bottles (Dewar flasks) with a copper-constantan thermocouple in each, one containing water with animals, the other water without animals. A thermocouple converts any slight temperature difference into an electrical signal. The output was read on a sensitive galvanometer, the deflection of the mirror indicating temperature difference between the two vessels. The earthworms (probably *Lumbricus terrestris*) produced 0.21 cal/hr per g at 15.6° C and 0.26 cal/hr per g at 19.5° C. A 4° increase in temperature therefore produced a 24 percent increase in metabolic rate. Except for the few with mechanisms for controlling body temperature within a narrow range, animals generally increase their oxygen consumption with increases in environmental temperature, often about double with a 10° rise. The earthworm result corresponds to a threefold increase in heat production for a 10° rise, which is not extraordinary.

The first rate of heat production by an adult marine invertebrate came from the gastropod *Nucella lapillus*, used by Grainger (2) in an apparatus similar to that of Hill (4). Six individuals together in the calorimeter produced 0.08 cal/hr per mg N. Snails of 7.5g total weight probably contained about 1.75g of living tissue, based on proportions of similar species. This makes 1 mg N equivalent to 122 mg tissue, and converts the rate to 0.66 cal/hr per g tissue (Table 6–1). The oxygen consumption of *Nucella*, sometimes called the "dog winkle," was determined separately as 16.8 μl/hr per mg N or 138 μl/hr per g tissue. The ratio of the two rates Q_H/Q_{O_2} is 4.76 cal/ml O_2, within 5 percent of the values for direct oxidation of common food substances, such as palmitic acid (4.65) and glucose (5.01). This result reconfirms the parallel between catabolism within cells and burning of foods in a calorimeter. In addition, it shows

TABLE 6-1.   Metabolic rates of invertebrates other than insects, determined by calorimetry.

	Weight (g)	Temperature (°C)	Q_H (cal/hr per g)	Reference
Annelida				
"worms"	(10)	19.5	0.26	4
Lumbriculus sp.	(0.01)	20.4	1.50	2
Crustacea				
Asellus aquaticus (Fresh water isopod)	0.107	20.4	0.53	2
Carcinus maenas (marine crab)	25.0	20	0.25	10
Cyclops abyssorum (Fresh water copepod)	0.0002	25	3.83	6
Mollusca (Gastropoda)				
Limnaea pereger (Fresh water)	(2.0)	20.4	0.42[x]	2
Nucella lapillus (marine)	7.47	20.4	0.66[x]	2
Biomphalaria glabrata (Fresh water)	0.75	27.0	0.64	1
Echinodermata				
Strongylocentrotus lividus (sea urchin blastulae)	- -	19.0	1.10[+]	8

Weight is mean total body weight of living animal. Those in parentheses are order-of-magnitude estimates, since actual weights were not given by authors.

Calculated using the following ratios:

[x] 1 mg N = 122 mg tissue

[+] 1 mg N = 62 mg tissue.

that a marine invertebrate is a source of metabolic heat, just as larger and warmer animals are, and suggests an alternate mode of measurement in cases where oxygen consumption may prove impractical.

Rates of heat production

The heat production of several aquatic invertebrates (Table 6-1) ranges from 0.25 to 3.83 cal/hr per g tissue. As expected, the larger animals, such as the shore crab *Carcinus maenas*, have lower weight-specific rates, and the smaller animals, such as the copepod *Cyclops abyssorum*, have higher rates. The lack of precise weights makes it impossible to establish any exact relation of heat production to body weight in these animals.

Heat production was preferred over oxygen consumption as a measure of total metabolism of *Carcinus maenas* subjected to variations in salinity because of the well-known but often inexplicable effect of salinity change on oxygen consumption (10). Animals of 15 to 35 g total weight produced heat at rates of 0.80 cal/hr per g at 20° C when first placed in the vessel, then at 0.40 after becoming quiet, and eventually at a rate of 0.25 cal/hr per g. This shows that activity may produce metabolic rates three times those in quiescent animals.

Two of these species, *Asellus* and *Nucella*, had oxygen consumption rates typical of combustion of foods, as described above. The snail *Limnaea pereger* has less heat production than predicted from oxygen uptake, a ratio of 2.60 cal/ml O_2. Since the two measurements were not made simultaneously, the oxygen uptake may have been abnormally high at the time, because of greater activity or repayment of oxygen debt. On the other hand, the small annelid *Lumbriculus* produced more heat than predicted, 11.26 cal/ml, possibly due to "some anaerobic breakdown of substrates" (2).

Most of the results discussed above were obtained with calorimeters constructed to retain heat and allow the temperature of medium and animal to rise. The work of Lamprecht (1,6), however, was performed with a different type of instrument. In 1948, Calvet described a microcalorimeter, based on the design principle of Tian, which allows the heat to escape from the experimental vessel and pass into a "heat sink," usually a block of aluminum. On the way, heat passes and activates thousands of small thermocouples, which provide output to a recorder, yielding a "thermogram." The thermogram continuously records power output in milliwatts.

Conversion to joules

In order to compare results of the various types of calorimetry, a common unit must be chosen. The calorie is a unit "not compatible with the SI (Système International), nor any other metric system," and therefore it may be expected to disappear gradually from the literature. The unit of heat is the joule (J). Thermochemical papers report ΔH in kJ per mole. The relation between these units is:

   1 joule/sec = 1 watt = 0.2389 cal/sec. Older values in the literature can be converted by means of: 1 cal = 4.1858J.

The matching of calorimetry against respirometry has great promise for the study of animals with the ability to shift from aerobic to anaerobic states. In recent years, more and more species from various groups have been shown capable not only of enduring one to four days of anoxia, but also ceasing to consume oxygen at low oxygen concentrations.

Bivalve metabolism during anoxia

Aerobic pathways of glucose oxidation capture chemical bond energy in the form of ATP with a relatively high efficiency, by the process of oxidative phosphorylation (Chapter 5). Studies on anaerobic glucose degradation to lactic acid in vertebrate muscle suggest a much lower efficiency. Therefore, animals that undergo periods of anoxia are believed to greatly reduce their "energy demand." Calorimetry measures total metabolism whether aerobic or anaerobic. Relations between total metabolism and aerobic metabolism alone are shown by the ratio Q_H/Q_O, which is generally expected to lie near values of $\Delta H/\Delta O$ for combustion of food substances. As mentioned earlier, this ratio varies from 4.5 to 6.0 cal/ml. In terms of SI units and good chemical practice, these are equivalent to 0.43 to 0.49 J/μmole. In the combustion of glucose, one mole requires six moles of oxygen, and the heat released is 2,817 kJ, giving $\Delta H/\Delta O$ of 0.470. Values of Q_H/Q_O greater than 0.49 indicate that total metabolism is exceeding oxygen consumption and that anaerobic processes are becoming important.

Marine bivalve mollusks often close their valves tightly and consume no oxygen at all for hours. Some species survive five to fifty days anoxia at low temperatures. This extraordinary tolerance of anoxia is presumed attributable in part to an efficient fermentation of stored glycogen and in part to a reduction in total metabolism so that less ATP is needed. Glycogen stores are abundant, about 20 percent of dry weight, and rates of glycogen disappearance are low, about 5 mg per day per g tissue. At this rate, glycogen lasts eight days. However, individual variation is so great that neither glycogen disappearance nor accumulation of acid end products are satisfactory as measures of anaerobic metabolic rates. When the valves are closed, muscular contraction, ciliary activity, heart beat, and biosynthesis of larger molecules are all decreased.

The ATP concentration in the tissues, and the balance between ATP and other nucleotides, have been used to estimate how much "energy demand" is reduced when bivalves shift from aerobic to anaerobic metabolism. No matter how well it is done, however, a chemical inventory cannot reveal the changes in rates that accompany transition to a new steady state. A better attack on the problem is to determine heat production before, during, and after the transition. Some efforts in this direction have been published recently (3,9). Both heat production and oxygen uptake of four species of bivalves are listed in Table 6-2. The ratio Q_H/Q_O was 0.462 J/μmole for *Mya arenaria*, in good agreement with the ratio $\Delta H/\Delta O$ for combustion of food substances. In three other species of bivalves, however, heat production indicated greater metabolic rates than the maximum oxygen consumption observed. This suggests that metabolism in some species, such as the oyster *Crassostrea virginica* with $Q_H/Q_O = 3.40$, is partially anaerobic most of the time. Tracer studies with ^{14}C-labeled substrates have already indicated that

TABLE 6-2. Metabolic rates of marine invertebrates, determined by calorimetry. Q_H in J/hr per g tissue, Q_O in μmole/hr per g.

	Weight (g)	Temperature (°C)	Q_H	Q_O	Q_H/Q_O
Crustacea					
Carcinus maenas*	25	20.0	1.05	– –	– –
Libinia dubia	8.4	24.6	1.24	2.61	0.477
Mollusca					
GASTROPODA					
Nucella lapillus*	7.47	20.4	2.75	6.17	0.446
Littorina littorea	4.14	22.4	1.62	3.82	0.424
Littorina irrorata[+]	1.75		0.078	0.170	0.456
PELECYPODA					
Mya arenaria	51.9	24.9	1.90	4.11	0.462
Mytilus edulis	5.32	23.9	1.59	2.50	0.635
Mercenaria mercenaria	91.6	21.5	3.07	2.68	1.145
Crassostrea virginica	18.6	24.1	6.23	1.83	3.40
Echinodermata					
Strongylocentrotus lividus*		19.0	4.62	11.82	0.390

*Recalculated from Table 1.

[+]From Pamatmat, 1978, animals in air under very low oxygen tension.
All other data from Hammen, 1979.

formation of succinic acid and other glycolytic end products occurs to some extent during aerobic metabolism. In time, the development of better methods will permit more precise determination of the relation between the various metabolic rates, not only in bivalves but in other anoxia tolerant forms such as polychaetes, sipunculids, brachiopods, and sea stars.

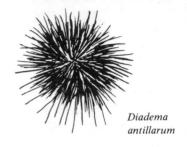

Diadema
antillarum

CHAPTER 7 Nitrogen excretion

Keeping animals healthy in a marine aquarium with fixed water supply depends on preventing toxic accumulation of waste products such as ammonia. All aquatic animals excrete ammonia as a major end product of protein and nucleic acid metabolism. Other products found in various lesser quantities include urea, uric acid, other purines, and free amino acids (FAA). Sometimes all these substances are grouped together as nonprotein nitrogen (NPN).

The Kjeldahl method

Rather than collecting urine from a marine invertebrate, which is often a difficult technical problem, one can simply hold the animal for a definite period in a small volume of seawater, and then analyze water samples. The relative fractions of the various products, sometimes called the nitrogen "partition," may be illustrated by some determinations made on the American oyster, *Crassostrea virginica* (Table 7-1). These data show that ammonia (NH_3) accounted for 65.0 percent of NPN in one case, and 68.7 percent of the four identified products in the

TABLE 7-1.   End products of nitrogen metabolism in excretion water of
100-g oysters, *Crassostrea virginica*. Rates in μmoles/day
per g tissue, as ammonia.

	Rhode Island May 1965		North Carolina July 1967	
	rate	percent	rate	percent
ammonia	1.562	65.0	0.644	68.7
urea	0.318	13.2	0.072	7.7
FAA	0.126	5.2	0.194	20.7
uric acid	nd*	—	0.028	3.0
sum	2.006	83.4	0.938	100.0
NPN	2.406	100.0	nd*	—

*not determined
(From Hammen, Miller, and Geer, 1966; and Hammen, 1968)

other case. Small quantities of urea, uric acid, and FAA were
also found, and about 15 percent of NPN remained unidentified.
NPN is determined by adding perchloric or trichloroacetic acid
to the water to precipitate any protein that might be present,
then performing a Kjeldahl digestion and distillation on the
filtrate. This old, reliable method converts all nitrogen-containing
substances to soluble ammonium sulfate by a selenium-catalyzed
digestion:

$$[2NH_3] + H_2SO_4 \xrightarrow{Na_2SeO_3} (NH_4)_2SO_4$$

Then strong alkali releases ammonia, which is distilled off,
captured in a measured quantity of weak acid such as boric
acid, and titrated:

$$(NH_4)_2SO_4 + 2NaOH \rightarrow 2Na^+ + SO_4^= + 2H_2O + 2NH_3^\uparrow$$

Practice makes perfect, so the total NPN and the sum of identified products approach equality as methods of analysis
improve. The number of possible substances that could be
excreted is so great, however, that a fraction of "unknown"
never disappears.

Origin of NPN

The principal source of excreted NPN is catabolism of amino acids, which come from hydrolysis of ingested protein and reside temporarily in intracellular "pools." Some fraction of the FAA pool is continuously withdrawn to be deaminated, yielding ammonia, and another fraction goes to form new species-specific proteins, not only during growth, but also to replace protein lost in the normal turnover of body constituents. The net result is that excretion may wax and wane, but it never ceases. Another source of NPN is catabolism of nucleic acids, which similarly deposit and withdraw from nucleotide pools. The bases adenine and cytosine have amino groups readily removable by appropriate hydrolytic enzymes.

Excretion of ammonia

Ammonia concentrations of $1-2\,mM$ are common in the blood of marine invertebrates (3). Higher concentrations are presumably toxic because they perturb acid-base balance with too much alkalinity:

$$NH_3 + H_2O \rightleftharpoons NH_4^+ + OH^-$$

Ammonia is very soluble, up to $5.3\,M$ even in cold water, and it diffuses freely across membranes. Much of its excretion, therefore, does not depend on kidneys, but occurs *via* the gills of crustaceans, for example, or across general body surfaces of creatures lacking an exoskeleton.

Excretion rates

Ammonia can be used for synthetic purposes, such as the production of glutamate from a-ketoglutarate, and the production of carbamyl phosphate from CO_2 and ATP, but there is always a surplus to be excreted.

Nitrogen excretion is a process much like gaseous exchange, in that it is a gross manifestation of cellular metabolism. Rates are expressed on the basis of tissue weight, as are oxygen consumption rates. Since the products differ in the number of N atoms per molecule and since the Kjeldahl method reduces all N to ammonia, it is best to give results as μmoles NH_3 in unit time per g tissue. The relation between weight of nitrogen and moles is:

$$\frac{1.0 \text{ g N}}{14.0 \text{ g N per mole } NH_3} = 0.0714$$

1 mg N = 71.4μmoles NH_3

Rates of excretion of four major products by six species of bivalve mollusks are given in Table 7-2. These animals range in size from *Solemya* with a total body weight of only a tenth of a gram to the oyster of more than 100 grams.

If N excretion is a standard, predictable feature of gross metabolism, like oxygen consumption, then rates may follow the surface rule $y = KW^{2/3}$ (Chapter 2). If weight-specific rate is used, the relation becomes:

$$y/W = \frac{KW^{2/3}}{W}$$

and

$$Q_N = y/W = KW^{-1/3}$$

This predicts that the ratio between excretion rate (Q_N) and $W^{-1/3}$ is a constant (K). The data in Table 7-2 agree with this hypothesis. Five species out of six yield values for K between 3.67 and 4.37, a range of variation less than 10 percent of the mean, 4.06.

Rates of ammonia excretion

The main component of excreted NPN in most marine invertebrates is ammonia. In the bivalve mollusks just considered, ammonia made up 50 to 75 percent of identified products. Rates

TABLE 7-2. Tissue-weight-specific rates of nitrogen excretion proportional to logarithm of total body weight in six marine bivalve mollusks ($y = kW^{-1/3}$). Mean total weight (W) in grams. Excretion rate (y) in μmoles/day (as ammonia) per g tissue.

Species	W	$W^{-1/3}$	y	k
Solemya velum	0.127	1.989	8.376	4.21
Donax variabilis	0.265	1.557	6.200	3.98
Modiolus demissus	4.00	0.630	2.313	3.67
Tagelus plebius	23.9	0.347	1.411	4.06
Mercenaria mercenaria	51.0*	0.270	0.695	(2.58)
Crassostrea virginica	101.9	0.214	0.936	4.37

*Young animal, does not conform to rule.
(From Hammen, 1968)

of ammonia release are influenced by the length of time that an animal is held in a small volume of water. For example, the ammonia excretion of the brachiopod *Lingula reevi* diminished to one-half its initial value after twelve hours, and to one-third after forty-eight hours (7). The effect was not due to starvation, because the rate was maintained constant for twelve hours when the water was changed every three hours. Accumulated ammonia evidently caused the animal to reduce its excretion rate. This type of stress can also affect the relative fractions of different end products. For example, the razor clam *Tagelus plebius* released a greater fraction of free amino acids and smaller fraction of ammonia and uric acid with duration of confinement up to twenty hours (Figure 7-1).

Osmotic stress also affects ammonia excretion rates. For example, release of ammonia by the shore crab *Carcinus maenas* diminished after twelve hours to one-third of its initial rate in seawater, but it remained relatively constant near 0.5μmole/hr per g in half-strength seawater (see Chapter 1). In studies of N excretion, care must be taken to keep salinity constant.

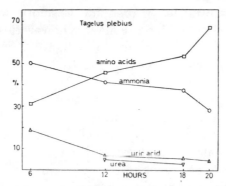

Figure 7-1. Proportions of nitrogen end products in excretion water of razor clam *Tagelus plebius* vary with duration of holding in small volume of seawater. (From Hammen, 1968)

Amino acid loss

Free amino acids make up substantial fractions of excreted NPN from many marine invertebrates. For example, the sea urchin *Diadema antillarum* released 61 to 65 percent ammonia and 26 to 29 percent FAA, and no urea, uric acid, or other purines (6). FAA lost to the medium are probably not waste products of metabolism, but represent an unavoidable leakage due to high concentrations within the tissues and cell membranes permeable to many small molecules. Table 7-3 shows that tissue FAA concentrations amount to 21 to 40 percent of

T A B L E 7-3. Concentrations of free amino acids in tissues of marine invertebrates (m*M*)

	tissue FAA	blood FAA	total osmotic strength (TOS) of medium	tissue FAA / TOS
Nereis virens	223	140	945	0.236
Mytilus edulis	366	—	1180	0.310
Crassostrea virginica	164	—	785	0.209
Crassostrea angulata	441	—	1188	0.371
Carcinus maenas	400	12	1000	0.400

total osmotic substances of mediums isosmotic with the tissues in animals from three phyla. FAA loss however, is not necessarily proportional to tissue concentration, as shown in Table 7-4. Bivalve mollusks had FAA loss equal to 23 to 38 percent of the sum of ammonia and FAA, while in *Carcinus* it was only 11.3 percent, in *Lingula* only 5.6 percent, and in *Nereis* there was no detectable FAA loss. These differences probably reflect differences in cell membrane permeability, as discussed in Chapter 1 on osmotic balance. Uptake of FAA is still another factor. For example, *Nereis virens* removed glycine from 2.0 mM solution most rapidly of any representative of ten phyla; lamellibranchs were also quite effective, but five crustaceans failed to take up glycine (10). Thus the amount of FAA found in the medium in any excretion experiment is the net result of uptake and loss. In *Nereis*, uptake predominates; in bivalve mollusks, leakage is greater. The FAA content of natural seawater varies with ecological conditions, with higher values reported from estuarine situations. For example, concentrations up to 2.8 μmoles/l were found in the York River (Virginia) (11), but a maximum of 0.27 μmole/l in the Irish Sea (9). In both studies glycine, alanine, and serine were among the five most abundant; the other two were aspartic and glutamic acids in the York River, and threonine and valine in the Irish Sea. Possibly compounds with lower molecular weight are more likely to leak out through cell membranes. Uptake by one species of FAA released

TABLE 7-4. Rates of ammonia excretion and amino acid loss by marine invertebrates (μmole/hr per g).

	Total weight (g)	Hours	NH$_3$	FAA	$\dfrac{\text{FAA}}{\text{sum}}$
Nereis virens	6.7	5	0.140	0	0
Modiolus demissus	4.0	24	0.063	0.033	0.344
Tagelus plebius	24.0	6	0.036	0.022	0.383
Crassostrea virginica	100.0	24	0.027	0.008	0.232
Lingula reevi	3.6	24	0.136	0.008	0.056
Carcinus maenas	30.0	6	0.260	0.033	0.113

by another in the same environment is a definite possibility that should be investigated.

Arginase and the ornithine–urea cycle

Urea and uric acid make up minor fractions of NPN. For example, three out of seven species of bivalves released 3 to 8 percent urea, and five out of seven released 1 to 7 percent uric acid (4). Usually the amount of urea excreted by a marine invertebrate is so small that it can be explained on the basis of degradation of dietary arginine alone, and it is not necessary to postulate a system of enzymes capable of generating urea. Arginase (EC 3.5.3.1.) activity has been found in tissues of many species:

$$
\begin{array}{ccccc}
\text{NH}_2 & & & & \\
| & & & & \\
\text{C} = \text{NH} & & & & \\
| & & & & \\
\text{NH} & & \text{NH}_2 & & \\
| & & | & & \\
\text{CH}_2 & \text{H}_2\text{O} & \text{CH}_2 & & \text{NH}_2 \\
| & & | & & \diagup \\
\text{CH}_2 & \rightleftharpoons & \text{CH}_2 & + & \text{C} = \text{O} \\
| & \text{arginase} & | & & \diagdown \\
\text{CH}_2 & & \text{CH}_2 & & \text{NH}_2 \\
| & & | & & \\
\text{HC} - \text{NH}_2 & & \text{HC} - \text{NH}_2 & & \text{urea} \\
| & & | & & \\
\text{COOH} & & \text{COOH} & & \\
\text{arginine} & & \text{ornithine} & &
\end{array}
$$

In the complete urea cycle, ornithine is combined with carbamyl phosphate (CAP) to form citrulline. The enzyme that catalyzes the production of CAP is called ATP: carbamate phosphotransferase or carbamate kinase (EC 2.7.2.2):

$$ATP + NH_3 + CO_2 \rightleftharpoons ADP + H_2N-C\underset{\textstyle O - PO_3H_2}{\overset{\textstyle O}{\diagdown}}$$

This reaction is extremely slow or nonexistent in all marine invertebrates examined so far (12), although the other enzymes of the urea cycle are often present. In order to demonstrate a functional cycle, isotope-labeled precursors are supplied to a tissue preparation along with an inhibitor of urease activity, to keep the animal's own urease from breaking down urea as rapidly as it is formed. One such inhibitor, highly effective at 10 mM, is acetohydroxamate (AHA):

$$\underset{\textstyle OH}{\overset{\textstyle CH_3}{C = NOH}}$$

Experiments with the lugworm *Arenicola cristata* showed urea formation from ^{14}C-citrulline and ^{14}C-arginine in the presence of AHA, but little or none from NaH^{14}CO$_3$. This indicates very low carbamate kinase activity. Since good activity has been found in several terrestrial invertebrates, it appears premature to conclude that no marine invertebrate can synthesize CAP. In some species, ammonia may enter into formation of CAP as the amino group of glutamine in addition to entering as ammonia itself (1). A second mole of ammonia enters the urea cycle as the amino group of aspartate. The original function of the cycle was probably nutritional, as arginine is required for synthesis of all proteins and the muscle phosphagen, N-phosphoryl arginine (Chapter 10); arginine can also contribute to osmotic adjustment. These functions may be more important than urea formation in marine invertebrates.

If an animal must lose nitrogen, the best answer is ammonia, since other products involve discarding carbon chains and their bond energy. If ammonia must be avoided, the next best product is urea, since it has the least C per N atom:

$$NH_3 \quad (NH_2)_2CO \quad C_5H_3O_3N_4 \quad C_6H_{14}O_2N_4$$

ammonia   urea   uric acid   arginine

Uric acid excretion and storage

Uric acid (2,6,8-oxypurine) is a highly oxidized degradation product of adenine and guanine, the purines occurring in nucleic acids. In most species the amount of uric acid excreted, 1 to 5 percent of NPN, is in the same range as the amount of nucleic acid in foods. The only group with exceptionally large uric acid production is gastropod mollusks. This production does not appear as an excretion but rather as accumulation in the tissues (1). The adaptive value of uric acid storage is unknown, but it seems to be related to the amount of time a gastropod spends out of water. Uric acid is soluble only to 0.15mM in water, so perhaps storage is less demanding of energy than circulating the large amounts of water needed to dissolve it.

The total lack of uric acid among the excretion products of some species probably indicates that they are capable of dismantling the purine ring through allantoin and allantoic acid to glyoxylate and urea. The entire purinolytic system has been detected in the bivalves *Mytilus edulis* and *Meretrix meretrix* (2). Of eleven species of polychaetes examined, all had measurable uricase activity, four had detectable allantoinase, two had traces of allantoicase, and four had urease activity; two species, *Hyalinoecia bilineata* and *Arenicola marina*, displayed evidence of all four enzymes (8). The purinolytic pathway, like arginase activity, is probably in many more species than the limited evidence indicates. Since both pathways produce urea, the total lack of urea excretion by many species requires explanation. It is probable that all marine invertebrates produce some urea, and that the amount they release to the medium is regulated by the activity of urease, also universal:

$$\begin{matrix} NH_2 \\ \diagup \\ C = O \\ \diagdown \\ NH_2 \end{matrix} \quad \xrightarrow[\text{urease}]{H_2O} \quad 2\,NH_3 + CO_2$$

The action of urease is the final step to ammonia in any pathway that includes urea. Varying the proportions of ammonia and urea in the NPN may be a means of adjusting acid-base balance (1). Ammonia produced by urease action could cause a shift from bicarbonate to carbonate ion:

$$NH_3 + HCO_3^- \rightleftharpoons NH_4^+ + CO_3^=$$

Increased urease activity would increase alkalinity, and thereby promote deposition of $CaCO_3$ in shell-forming animals (See Chapter 8). Rates and patterns of nitrogen excretion, through their influence on pH, have the potential of affecting almost every other aspect of chemical physiology.

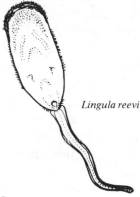

Lingula reevi

CHAPTER 8 Shell formation

Calcareous shells are produced by brachiopods, bryozoans, crustaceans, mollusks, and some polychaetes. They serve as protective shields against predation. Echinoderms have calcareous plates embedded in their tissues and external calcareous spines for protection. Shell growth is a conspicuous biosynthetic activity that continues throughout the life of many species. Most animals periodically or continuously deposit new, larger layers of shell material, and even the crustaceans reabsorb some of the calcium in their integument when molting and recycle it in constructing a new exoskeleton.

Shell composition

Shells consist of various combinations of mineral and organic material. The most common mineral is calcium carbonate ($CaCO_3$). The principal crystalline form is calcite, but aragonite is also common. Much less common is vaterite, "a precursor of the more stable polymorphs, aragonite and calcite, in the repair of exoskeletons of adult gastropods and bivalves" (5). Aragonite is denser and harder than calcite and belongs to a different system (Table 8-1). The hexagonal system consists of crystals that can be referred to a principal axis and three equal lateral axes perpendicular to it and 60° apart. The orthorhombic system contains crystals that are defined by three unequal perpendicular axes.

Some species of mollusks have shells consisting almost exclusively of calcite (*Crassostrea virginica*), others entirely aragonite (*Mercenaria mercenaria*), and still others normally have both forms in separate layers (*Mytilus edulis*). Deposition of vaterite occurs during repair of damaged shells.

Shell minerals are not chemically pure. As shown in Table 8-1, small amounts of strontium and magnesium, elements heavier and lighter than Ca in the same column of the periodic table, are regularly present in biological calcite and aragonite. The ratio Sr/Ca in seawater, however, is 2.5 times greater than the ratio in molluscan shells, indicating a discrimination against Sr. The ratio Mg/Ca is 5000 times greater in seawater, where Mg is 54.5 m*M* and Ca 10.5 m*M*, indicating a virtual exclusion of Mg from shells. Mineralogy controls Mg content (only one-tenth as much in aragonite) and crystal growth kinetics and/or metabolism controls Fe, Sr, and Mn content (4).

The majority of living brachiopods have calcareous shells, but some of the inarticulates contain calcium phosphate rather than calcium carbonate.

The great variety of types of shell composition found in inarticulate brachiopods is shown in Table 8-2. The chitinophosphatic type has existed since well before the origin of

TABLE 8-1. Forms of calcium carbonate in bivalve mollusk shells.

	Calcite	Aragonite	Vaterite
Hardness (Mohs scale)	3.0	3.5–4.0	—
Specific Gravity (g/cm^3)	2.710	2.947	2.645
Crystal System	rhombohedral	orthorhombic	hexagonal
Strontium Content (percent)	0.15	0.20	?
Magnesium Content (percent)	0.20	0.02	?

TABLE 8-2. Composition of shells of representative genera of
inarticulate brachiopods, each of a separate family.

	Mineral	Organic
Crania	$CaCO_3$ (calcite)	protein, about 1%
Discinisca	$Ca_3 (PO_4)_2$	chitin and protein, 11%
Lingula	$Ca_3 (PO_4)_2$	chitin and protein, 42%
Obolella[1]	$CaCO_3$ (calcite)	unknown
Trimerella[2]	$CaCO_3$ (aragonite)	unknown

[1] extinct since Cambrian
[2] extinct since Silurian

vertebrate bone, which resembles it in mineral composition. The
fraction of organic material (42 percent) in *Lingula* shells is
much greater than found in carbonate shells (8).

Shell growth

Shell growth depends on precipitation of Ca^{++} and $CO_3^=$ from
solution onto the surface of shell already present. Living cells
form an epithelium that secretes the necessary ions into a liquid
phase lying between the tissue and the shell. In mollusks and
brachiopods, the organ bearing this specialized epithelium is
called the mantle. The mantle consists mainly of connective
tissues sandwiched between inner and outer sheets of epithelium. Usually no more than 1mm in thickness, it nevertheless
contains canals for blood and coelomic fluid, and muscle fibers
that enable the mantle to retract away from the margin of the
shell. The activity of the mantle is revealed in part by analyzing
the extrapallial fluid. The extrapallial fluid of three mollusks
had a Ca concentration 15 to 28 percent greater and a total
CO_2 concentration 68 to 108 percent greater than that of seawater (9). Other ions were not significantly different. The pH
was lower (7.3-7.4) than that of the seawater (7.9), probably
because of the enrichment of CO_2. Most of the carbon dioxide
of seawater is in the form of HCO_3^-. This must be converted to

$CO_3^=$ for precipitation. As mentioned in Chapter 7, ammonia may serve as the proton acceptor to achieve this conversion.

Rates of shell growth vary greatly with species and season. In the pearl oyster *Pinctada martensii*, the maximum rate was $0.75\,mg/day/cm^2$ in August when water temperature was highest. This rate would result in an increase in thickness of 1 mm per year. During several winter months, shell decalcification exceeded deposition, and the growth rate was actually less than zero (3). Prolonged shell closure also reverses mineral deposition, as metabolic acids accumulate and dissolve $CaCO_3$. Experimentally, low Ca in the seawater and inhibitors of urease and carbonic anhydrase also decreased the rate of mineral deposition (10).

In addition to concentrating Ca^{++} and $CO_3^=$ in the extrapallial fluid, the mantle also secretes matrix proteins, collectively called conchiolin, and the outer sheath protein, periostracum.

Organic matrix

Many analyses of the organic residue after decalcification of mollusk shells have shown that conchiolin consists of at least two proteins, a scleroprotein and a water-soluble protein (like gelatin). The most abundant amino acids in hydrolyzates of conchiolin are glycine, alanine, serine, aspartic acid, and glutamic acid. The collagens of marine invertebrates are rich in glycine (12.5 to 18.8 percent by weight) and relatively abundant in hydroxyproline (5.0 to 8.6 percent). Some conchiolins lack hydroxyproline, and some have x-ray diffraction patterns similar to β-keratin, the protein of hair, horn, and claws. Thus the scleroprotein fraction of conchiolin is similar to a collagen but is different enough that it deserves a name of its own. The term "nacroine" has been proposed (2). Glucosamine is present in some conchiolins. In others, especially from *Nautilus*, the glycoprotein consists in part of chitin, a polymer of N-acetylglucosamine. Chitin is a major component of arthropod exoskeletons. A high glycine content and small amounts of chitin also have been found in periostracum.

The mantle, therefore, synthesizes chitin and at least two kinds of structural protein. The extrapallial fluid of many mollusks contains proteins, but neither these proteins nor those of conchiolin have been characterized sufficiently to determine whether they are identical or not. A protein dissolved in the extrapallial fluid must undergo a transformation to become an insoluble protein of the shell matrix.

The deposition of shell starts with formation of very minute crystal nuclei on a sheet of conchiolin. The nuclei grow to the size of seed crystals (50 nm) visible by electron microscopy. As the crystal grows, soluble protein becomes incorporated in it, and when a layer is complete, it is covered by a new sheet of conchiolin. The microarchitecture varies with species.

Mantle metabolism

The work of shell formation depends on energy made available by the metabolism of the mantle. Mantle tissue of oysters consumes more oxygen than adductor muscle, but only about one-half that of digestive gland and gill (Table 8–3). Similarly,

TABLE 8–3. Oxygen consumption of isolated tissues of three species of oyster, genus *Crassostrea* (ml/hr per g dry).

Species	C. angulata	C. gigas*	C. virginica
Temperature (°C)	27.5	25	24
Digestive gland	1.95	2.46	—
Gill	1.72	2.155	2.76
Mantle	0.96	1.28	1.27
Adductor	0.25	—	0.24

*Calculated on assumption that tissue contains 80 percent water.
(From Chapheau, 1932; Mori, 1968; Percy et al.; 1971.)

mantle of four species of bivalves had only 60 percent of the oxidoreductase activity of muscle from the same animals (3). Thus, the work of the mantle, although crucial, either requires less energy than the work of other tissues or it obtains a greater fraction by anaerobic processes. A significant aspect of mollusk metabolism in relation to shell formation is the tendency to produce succinic acid rather than lactic acid during anaerobic periods. The ionization constant of lactic acid is double that of succinic acid (giving pKa of 3.86 and 4.19, respectively). Succinate production requires CO_2 fixation, and therefore acts in two ways to reduce acidity that would dissolve the shell: increase in carbonic acid is retarded and a weaker acid than lactic is produced.

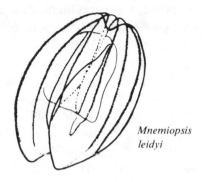

Mnemiopsis
leidyi

CHAPTER 9 Bioluminescence

In the Caribbean sea, the Mediterranean, or Narragansett Bay, any disturbance of the water at night produces flashes of light indicating the presence of previously invisible bioluminescent organisms. A single flash may be as brief as 0.2 millisecond or as long as 10 seconds, depending on what animal produced it. This can be determined only by catching and examining specimens. Luminescent species are found in thirteen of the thirty-three animal phyla (5), including every major group of coelenterates except corals (7). Among the most studied species are the clam *Pholas dactylus,* the ostracod crustacean *Cypridina hilgendorfii,* the ctenophore *Mnemiopsis leidyi,* the hydromedusa *Aequorea forskalea,* and the colonial hydroid *Renilla reniformis,* sometimes called the "sea pansy."

Luciferins and luciferases

The essential reaction in many species is the oxidation of a substrate, luciferin, by molecular oxygen (6). The enzyme, luciferase, catalyzes the production of oxyluciferin, a molecule with its first electronic singlet in an excited state. Light is emitted when oxyluciferin returns to the ground state. Carbon dioxide is split off as a result of a specific opening of the ring structure of luciferin.

Luciferins from *Cypridina* and *Renilla* (Figure 9-1 and Figure 9-2) are not identical, but have certain common features. They are polycyclic, heterocyclic compounds of relatively low molecular weights (*c.* 400). The central parts of their structures, the pyrazine nuclei, are identical, even to the keto group that makes them pyrazin-3-one derivatives. *Cypridina* luciferin has one side chain composed of an indole ring, like that of tryptophan, and another side chain consisting of a short carbon chain ending in a guanidino group, much like an arginine residue. *Renilla* luciferin has a phenyl ring and two p-hydroxyphenyl groups attached to the central pyrazine, the same groups that occur in phenylalanine and tyrosine. This structure seems common to all luminescent coelenterates, and has been called "coelenterazine" (8). The four amino acids just named probably serve as precursors in the synthesis of luciferins; experiments to test this hypothesis would be useful. The "firefly squid" *Watasenia scintillans* contains a sulfonated version of coelenterazine. Thus these luciferins are isologues, substances that play the same role in metabolism *and* show signs of chemical kinship (see Chapter 5). *Cypridina* releases minute quantities of both substrate and enzyme to the water, where the light-producing reaction occurs in the immediate vicinity of the animal. *Renilla* retains its spent luciferin and recycles it.

The enzyme luciferase, which catalyzes the combination of one mole of oxygen with one mole of luciferin, is a protein of molecular weight 50,000 in *Cypridina hilgendorfii.* It is an oxygenase and would be expected to appear in the list with twenty-seven other enzymes (EC 1.13.11.-) capable of incorporating two atoms of oxygen into products of the reactions. However, it is not listed in the 1973 edition of "Enzyme Nomenclature," probably because it was not sufficiently characterized at that time. No co-factor or prosthetic group is

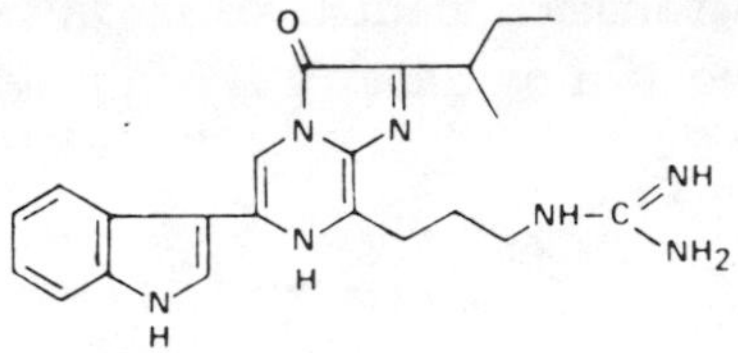

Figure 9-1. Cypridina luciferin, $C_{22}H_{27}N_7O$, mol wt 405, emission λ max = 460nm, blue. From Clayton, 1971.

known for *Cypridina* luciferase; the *Renilla* luciferase requires diphosphoadenosine (2).

Luciferase of the hemichordate *Balanoglossus* also catalyzes the oxidation of reduced luciferin (LH_2), but in this reaction the oxidant is hydrogen peroxide:

$$2\,LH_2 + H_2O_2 \rightarrow 2\,L + 2\,H_2O$$

This luciferase is therefore a peroxidase, like those that oxidize NADH, cytochrome *c*, and glutathione. Other peroxidases are capable of replacing *Balanoglossus* luciferase in the light-producing reaction (1). These are classified EC 1.11.1.-.

Sources of hydrogen peroxide

Hydrogen peroxide is a very reactive and potentially harmful substance that is normally produced in cells, but is not allowed

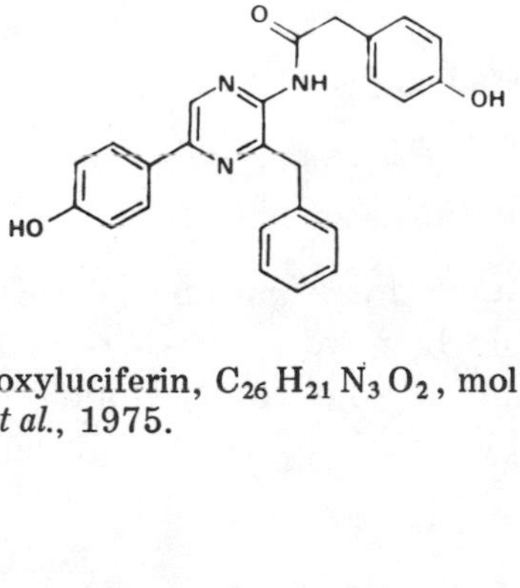

Figure 9-2. Renilla oxyluciferin, $C_{26}H_{21}N_3O_2$, mol wt 407, λ max = 435 nm. From Cormier *et al.*, 1975.

to accumulate to toxic concentrations. One source of H_2O_2 is the action of amino acid oxidases:

$$\underset{\text{amino acid}}{\overset{R}{\underset{|}{\underset{\text{COOH}}{\overset{|}{H_2N-C-H}}}}} + O_2 \xrightarrow{\text{ } H_2O \text{ }} \underset{\text{keto acid}}{\overset{R}{\underset{|}{\underset{\text{COOH}}{\overset{|}{C=O}}}}} + NH_3 + H_2O_2$$

Amino acid oxidases are not uncommon; for example, several mollusks have enzymes active on arginine, histidine, and lysine (3).

Superoxide dismutase

Another source of H_2O_2 is the dismutation of superoxide radicals. The superoxide radical arises directly from oxygen:

$$O_2 + e^- \rightarrow O_2^-$$

Formation of this radical is probably the reason for toxicity effects in all atmospheres with greater partial pressures of O_2 than that of air. Any system which generates O_2^- soon accumulates H_2O_2 (4), by the action of superoxide dismutase (EC 1.15.1.1):

$$O_2^- + O_2^- + 2H^+ \rightarrow H_2O_2 + O_2$$

Superoxide dismutase has been found in bovine erythrocytes, chicken liver cells, and cells of spinach, yeasts, and molds. Such a broad distribution suggests that it probably occurs in almost all organisms. Regardless of how H_2O_2 is produced, however, peroxidases are needed to dispose of it.

Detoxification hypothesis

Luciferases and luciferins seem special reagents in reactions forming a general biochemical cycle of peroxidation and cleavage, used by all organisms to detoxify some of the more harmful forms of oxygen (1). These reactions, normally dark, occasionally have requirements of spin and symmetry conservation that result in an "excited product." Light is emitted when this product returns to the ground state.

Photoproteins

Bioluminescence in most groups of marine invertebrates is consistent with the oxygen detoxification hypothesis. However, the hydromedusa *Aequorea* has a luminescent system consisting of a single conjugated protein called aequorin, which is not dependent on oxygen for activation. Crude preparations of aequorin emit light in response to a great variety of stimuli, such as heat, pressure, and organic solvents. Purified aequorin, however, is triggered only by Ca^{++}. The complex photoprotein has a chromophore identical to the luciferin of *Renilla* and a protein of molecular weight 30,000. Aequorin therefore resembles a permanent combination of luciferin and luciferase, ready to react upon contact with Ca^{++}. This luminescence is still somewhat consistent with the oxygen detoxification idea, because both oxygen and fresh coelenterazine are required to regenerate a previously reacted system.

Another case of closely linked enzyme and substrate is the photoprotein of the polychaete *Chaetopterus*. This has a high molecular weight (184,000), requires several co-factors (including Fe^{++}), and reacts only when both O_2 and H_2O_2 are present.

Chemical accident

Bioluminescence serves different purposes in different animal groups: species recognition, aid in finding a mate, attraction of prey, avoidance of predation. The reactions are also varied. Although the luciferins of marine invertebrates have proved remarkably alike, they are quite different from firefly and bacterial luciferins. The luciferases vary in molecular weight and degree of binding with the luciferins. Co-factor requirements range from very complex to simple to none.

The general opinion at present is that luminescence is a "chemical accident" that has been exploited for various biological purposes, rather than a fundamental trait. The need to detoxify some forms of oxygen is the most likely basic survival need that has been identified. This way of thinking helps to explain why light production occurs in similar ways in widely separated phyla and yet is lacking in some species closely related to luminescent species.

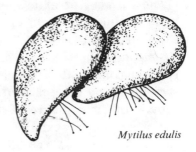

Mytilus edulis

CHAPTER 10 Ciliary activity
and muscular contraction

Sea anemones, and other anthozoans with tentacles too short to convey food into the mouth, manage to feed by means of cilia. These are very small, very numerous, hairlike structures that beat or bend in a coordinated fashion. The cilia of the anemone move small particles of food toward the gullet. The gills of many bivalve mollusks and many polychaetes are also covered with cilia, often arranged in definite tracts so as to form a kind of conveyor belt, moving food particles to the mouth. The lophophore of brachiopods, phoronids, and ectoprocts is a tentaculated feeding structure well supplied with tracts of cilia. The larvae of echinoderms employ cilia for both feeding and swimming (8), and so do larvae of polychaetes, echiuroids, mollusks, entoprocts, and hemichordates. The sponges are equipped with choanocytes lining their canals, cells with a beating flagellum to produce currents. A flagellum is basically a longer version of a cilium, both structures about $0.2\,\mu$m in diameter, cilia 10–$20\,\mu$m long, and flagella $20\,\mu$m to 2mm long (10).

Ultrastructure of cilia

The filament or axoneme (11) of a cilium originates from a basal granule within an epithelial cell, and it is enclosed by a sheath continuous with the cell membrane. Analysis of fine structure by electron microscopy shows the filament characteristically composed of a pair of central fibrils or microtubules surrounded by nine pairs of peripheral microtubules. The main protein of the structures is called *tubulin* (mol wt 60,000); each peripheral pair has extensions ("arms") made of another protein called *dynein* (mol wt 600,000); a protein linkage material between microtubules is called *nexin* (mol wt 165,000).

Rate of ciliary beating and work done

Under ordinary microscopy the cilia seem to beat rapidly, and in fact their angular velocity is of the order of ten strokes per second, through a semicircle. But the distances are relatively small, the tip of a flagellum or cilium actually moving through only about 1.1 mm/sec to 1.3 mm/sec or 4.0 m/hr to 4.7 m/hr. The effective stroke is often a swift lashing out of a relatively rigid cilium like the stroke of an oar, while the recovery stroke is slower and may involve much bending. A polychaete gill cilium required 20 msec to swing through a semicircle, then 40 msec to return to its original position, thus the whole cycle was occurring at 16.7 beats/sec (10).

The work done by a beating cilium can be calculated from its size, the viscosity of water, and the rate at which it is accelerated through the water. The cilium is so small that its own inertia is insignificant. The unit of force is the newton, which moves 1 kg with an acceleration of 1 m/sec² (f = ma), and the unit of work is the joule, which is a force of 1 newton operating over a distance of 1 m (w = f × d). A cilium does work at a rate of the order of 3×10^{-14} J/sec (2).

The rate of doing work is power, and the power of a cilium is therefore about 3×10^{-14} watt. This seems infinitesimal, but cilia are concentrated in tracts with a surface density of the order of 30 million per square mm, and the cells or particles that they have to move are very small. Thus a particle of $1 mm^2$ surface and $1 mm^3$ volume would have 30 million cilia working on it, with a total power of 1×10^{-6} watt = 1 microwatt, sufficient to impart an acceleration of $1 m/sec^2$ to the 1-mg particle. Ciliary feeders retain particles a few μm in diameter (8), so the actual areas, numbers of cilia, and power available are all much smaller, but proportionality allows movement over distances rarely greater than 1 cm in a few seconds.

Dynein

A clue to the riddle of the energy source of ciliary activity is the ATP-ase (EC 3.6.1.3 = adenosinetriphosphatase) activity of dynein:

$$ATP + H_2O \rightleftharpoons ADP + Pi$$

Dynein purified from cilia of a protozoan, *Tetrahymena pyriformis*, catalyzed ATP hydrolysis with a specific activity of 1.3–3.5μmoles phosphate/min per mg protein. With a molecular weight of 600,000, this activity is equivalent to 13–35 molecules ATP dephosphorylated per second per enzyme molecule.

Rates of ciliary and flagellar beating are of the same order of magnitude, 13–35 beats/sec. The hypothesis has been proposed that the passage of a bend along a cilium is associated with the use of one ATP molecule by each dynein molecule along that half of the cilium (2). A test of this hypothesis was performed with glycerinated spermatozoa of a sea urchin *Lytechinus pictus.* Many endogenous substances, including ATP, are leached out by soaking the cells in 50 percent glycerol for a few hours at low temperatures, but the dynein remains intact. Beat frequency increased from 5 to 25 beats/sec as the concentration of added ATP increased, from 0.1 to 3.0 m*M*. The same suspension of glycerinated cells was used in ATP-ase assays at the same pH and temperature. Enzymatic activity was proportional to beat frequency with an average rate of reaction 0.65×10^{-19} mole ATP per beat.

If ATP is the sole energy source, then the amount of chemical bond energy transduced into flagellar work can be calculated as:

$$\frac{3 \times 10^{-14} \, J/sec}{0.65 \times 10^{-19} \, mole/beat \times 30 \, beats/sec} = 1.54 \times 10^4 \, J/mole$$

The bond energy in the terminal phosphate of ATP is about $33 \, kJ/mole$. Thus there is more than enough energy to account for the necessary expenditure.

An ATP-dynein hypothesis

One can go a step further to test the hypothesis of a one-to-one ratio of dynein and ATP molecules, by estimating the number of dynein molecules that can be contained in a flagellum. Their distance apart in a *Tetrahymena* cilium is 16–20nm. Assuming equal spacing, a *Lytechinus* flagellum $42 \, \mu m$ long has 2330 dynein molecules end-to-end, arranged in nine double fibrils ($18 \times 2.33 \times 10^3 = 4.2 \times 10^4$ dynein molecules per flagellum). This result may be compared with the number of molecules of ATP dephosphorylated per beat of a single flagellum, which was estimated above as 0.65×10^{-19} mole/beat. Multiplying by Avogadro's number:

6.023×10^{23} molecules per mole $\times 0.65 \times 10^{-19}$ mole/beat
$= 3.91 \times 10^4$ molecules ATP/beat

This agrees with the hypothesis that each dynein molecule uses one ATP molecule during each beat cycle (2).

Molecular mechanism of ciliary action

Precisely how the chemical energy is transduced into mechanical work is not wholly explained, but one feature of the process is now fairly clear. The individual microtubular filaments do not shorten to cause bending, but rather slide past one another on one side of the cilium or flagellum, resulting in a bend toward that side (9).

Muscular contraction velocity

Compared with a flagellum whipping back and forth at unequal rates, beating sometimes in a spiral motion, and even reversing its direction at times, muscular contraction seems very simple. It isn't. Even the mechanics of shortening are still under debate.

As the load, or weight that it must lift is increased, the speed of contraction of a muscle diminishes, but not in direct proportion. In actual experiments, an isolated muscle is arranged on a lever that permits the same amount of (isometric) shortening with varied loads and does not permit stretching. The muscle is stimulated, usually by electric shock, and the velocity is measured from the trace of a smoked-drum kymograph or other recording device. The results are presented in a graph of velocity (v) in mm/sec $vs.$ load (P) in g/cm^2, as in Figure 10-1. The analysis consists in finding an equation that describes the curve of the data, which was done in 1938 by A.V. Hill (4). The Hill equation is:

$$v = \frac{(P_O - P)b}{P + a}$$

In addition to v and P, specified above, P_O is the maximum load that a muscle can lift under tetanizing stimulation, a is a parameter with dimensions of force, and b is a parameter with dimensions of velocity. Parameters are numbers chosen by the experimenter to make the equation fit the data; P and v are not parameters, but variables or factors.

P_O and P are expressed as weight per unit area because the work that a muscle can do clearly increases with its thickness and the number of fibers involved. P_O varies with species and with individual muscles, is usually of the order of $1-10\,kg/cm^2$,

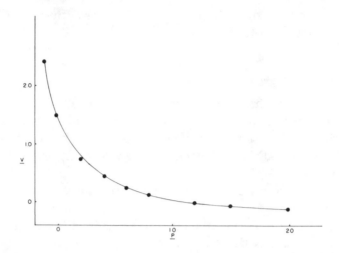

Figure 10-1. Relation between velocity of contraction (*v*) in mm/sec and load (P) in g/cm² for the byssus retractor of *Mytilus edulis*. Values for points representing *v* and P less than zero are calculated. Curve of the Hill equation corresponds closely to experimental points.

but occasionally lies outside this range. For example, the opaque part of the adductor of the oyster, *Crassostrea angulata*, has P_O = 7.6 kg/cm² (7) and the slow part of the remotor of the second antenna of the lobster, *Homarus americanus*, has P_O = 2.8 kg/cm² (6). Often the load is expressed as a fraction of the maximum (P/P_O), in which case all values lie between 0 and 1. This helps in drawing the graph, but conceals important information.

The Hill equation has the form of an hyperbola, $y = k/x$, but because of the parameters *a* and *b*, the asymptotes lie somewhere outside the upper right quadrant of Cartesian coordinates (*x*, *y* axes). For about 26 years, *a* appeared to be the heat of shortening per cm, so $(P + a)v$ was the total rate of energy liberation, as work and heat together. (Hill had received a Nobel prize in 1923 for work on heat production during contraction.) But this relation was "too good to be true" (4); at present no exact physical meaning can be ascribed to a. In magnitude a is usually 11 to 26 percent of P_O; a/P_O is 0.13 for cephalopod smooth muscle.

With P_O found by experiment and a and b found by curve fitting, the mechanical properties, with respect to force and velocity, are fully specified. The maximum velocity of contraction (v_O) occurs when $P = 0$. Some muscles must contract rapidly and repeatedly to perform their function; for example, the fast part of the antennal remotor of *Homarus*, responsible for sound production, has a high frequency, 100/sec, but a relatively weak force, $P_O = 0.025\,kg/cm^2$. Other muscles contract more slowly and maintain the contraction for long periods; the most extreme examples of this kind are the molluscan "catch" muscles, responsible for keeping valves closed for hours. These muscles have low v_O and high P_O; for example, the *Mytilus* byssus retractor $v_O = 1.5\,mm/sec$ and $P_O = 12\,kg/cm^2$, and the opaque adductor of *Crassostrea angulata* is similar in all essential features (7). Often values of b and v_O are given, for the sake of comparison, as fractions of the standard length (L_O); if the length of the *Mytilus* muscle is 6.0 mm, then $v_O/L_O = 0.25$ length/sec; the heart muscle of *Limulus* is four times more rapid, $v_O/L_O = 0.98$ length/sec.

Chemistry of contractile proteins

The principal structural proteins of muscle are *actin* (mol wt 48,000) and *myosin* (mol wt 500,000), analogous to the tubulin and dynein of cilia, especially since myosin is also an ATP-ase. In a cycle of muscular contraction, actin and myosin become bound in a complex called actomyosin, which catalyzes the dephosphorylation of ATP while releasing previously bound water. At the molecular level, the contraction consists of thin actin filaments sliding past thick myosin filaments by means of forming and breaking cross bridges between them. Two minor proteins, *troponin* (mol wt 80,000) and *tropomyosin* (mol wt

54,000) function in the relaxation process that follows a contraction or "twitch," as it is sometimes called. According to this sliding filament model, maximum tension (P_O) varies with length of myosin filaments, since this determines the amount of cross-bridge bonding, while velocity (v_O) varies inversely with length. The size of myosin filaments is generally 4–7 μm long by 16–50 nm in diameter in striated muscle and quite variable with species, while actin filaments are more uniform at one-fourth the diameter of myosin filaments. In molluscan catch muscles, the low v_O and high P_O have been correlated with extraordinarily long filaments (up to 50 μm × 100 nm) built up from myosin on a paramyosin core (3).

The outline of the contraction cycle given above has been embellished by significant details. For example, Mg^{++} must be supplied in equimolar concentration with ATP in systems where myosin acts as an ATP-ase. And in intact muscle fibers, Ca^{++} sequestered in sarcoplasmic reticulum of resting muscle is liberated into the sacroplasm to initiate contraction.

Barnacle muscle fibers

The giant muscle fibers of a barnacle, *Balanus nubilis*, have proved highly useful for investigating contractile events on the cellular level. The scutal-tergal adductor is made up of about twenty-five cross-striated fibers from 0.5 to 2.0 mm thick (5). With tetanus tensions of 50 g per fiber, P_O = 5.0 kg/cm². The great advantage of giant fibers is that intracellular electrodes can be inserted to measure trans-membrane potentials, and minute syringes can be used to inject various interesting substances without significant damage to the fiber. By these techniques, a mean resting potential of –50 mV was found in the *Balanus* muscle cell, and action potentials from 15 mV to 40 mV, increasing with the size of the twitch. With a constant stimulus strength of 3.0 volts, the increase of stimulus duration from 150 to 300 msec brought increasing Ca^{++} release, increasing membrance depolarization, and increasing isometric tension (1).

The release of previously bound Ca^{++} from crustacean muscle fibers has been detected by three methods, all dependent on micro-injections. Radioactive ^{45}Ca has been used as a tracer of calcium movement; a chelating agent, EGTA, binds Ca^{++} and inhibits contraction in proportion to injected dose; and most ingenious of all, a purified photo-protein, aequorin from a hydromedusa, gives an intracellular flash of light when it combines with Ca^{++} released during contraction!

Muscle phosphagens

The ATP used in muscular contraction must be continuously replenished, and the immediate source is the phosphagen kinase reaction, wherein phosphagens such as N-phosphoryl arginine (PA) donate phosphate to adenosine diphosphate (ADP):

PA + ADP → arginine + ATP

The reaction is catalyzed by arginine kinase (EC 2.7.3.3).

$$HN \sim PO_3H_2 \qquad\qquad\qquad HN \sim PO_3H_2$$
$$/ \qquad\qquad\qquad\qquad\qquad /$$
$$HN = C \qquad\qquad\qquad\qquad HN = C$$
$$\backslash \qquad\qquad\qquad\qquad\qquad \backslash$$
$$HN-CH_2-CH_2-CH_2-CH-COOH \qquad HN-CH_2-COOH$$
$$| \qquad\qquad\qquad\qquad |$$
$$NH_2 \qquad\qquad\qquad CH_3$$

N-phosphoryl arginine   N-phosphoryl creatine

In some animals, PA is not the phosphate storage compound; there are five other guanidino compounds that sometimes serve the same function, the most common after PA is N-phosphoryl-creatine (PC). Some sponges and some species in three classes of echinoderms, the echinoids, holothuroids, and ophiuroids have both PA and PC, while asteroids and crinoids have only PA. Four different kinases are found alone and in various combinations in various species of polychaetes, which suggests that the ancestral gene for the synthesis of arginine kinase has undergone duplication, mutation, loss, and further mutation (12). Thus the chemistry of muscular contraction is closely related to biochemical evolution. The connection of these subjects was suggested in Chapter 5 in the discussion of lactate dehydrogenase specificity of various taxonomic groups. Many animals, of course, rely on lactate dehydrogenase to generate ATP under anaerobic conditions, such as arise in vigorously contracting muscles.

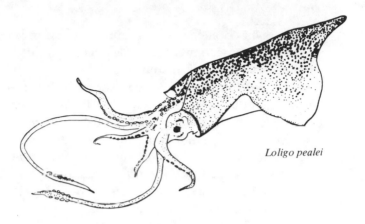

Loligo pealei

CHAPTER 11 Nervous conduction

Moving at a leisurely pace through the water, a squid, threatened by a predator, departs at high speed. The entire mantle musculature contracts at once, forcing water in the mantle cavity out through a narrow passage of the funnel, so that the mollusk moves away by jet propulsion. This is a rapid retreat; the funnel is directed forward like the extended tentacles, although it can be used to advance also. Speeds up to 20 knots have been reported. Relaxation of the circular muscles allows a new supply of water to enter.

The giant axon system

The signal for mantle contraction originates in two neurons in the brain, in response to sensory information. The signal goes to a visceral pallial ganglion which controls the muscles of the funnel and those connecting mantle and head. It then goes to a stellate ganglion, which controls the mantle muscles. The stellate ganglion contains thousands of nerve cell bodies; the axons of groups of 300 to 1500 cells fuse to produce single axons 0.65 to 1 mm in diameter and 5 cm long, which go out in six or eight branches to terminate on muscles. This "giant axon" system accounts for the extreme rapidity of conduction of the nerve impulse in squids, such as *Loligo pealei*, the common United States Atlantic coast species. Invertebrate nerves in general conduct impulses at 0.1–1.0 m/sec., while the *Loligo* stellar nerve has a conduction velocity of 20 m/sec.

Giant axons are not limited to cephalopod mollusks. They are found in polychaetes, such as *Nereis virens*, 0.04 mm diameter, and *Diopatra cuprea*, 0.10 mm, with conduction velocities of 5 m/sec and 10 m/sec, respectively; also in crustaceans such as *Palaemon serratus*, 0.05 mm and *Homarus americanus*, 0.16 mm, with velocities of 23 m/sec and 18 m/sec. Velocity is roughly proportional to diameter of axon, but is greatly increased by the presence of a myelin sheath, present in *Palaemon*, but lacking in the majority of marine invertebrates. In some animals, polychaetes, nemerteans, echinoderms, *Limulus*, there is a neuropile or nerve cord consisting of many small neurons whose axons extend together through one or two segments, then synapse with others. In coelenterates, there are nerve nets of individual but connected neurons.

If sponges have a nervous system, it is very simple and lacks typical neurons. Conduction velocities in neuropiles are 0.02–0.54 m/sec, in nerve nets, 0.04–0.90 m/sec. A "nerve" generally consists of an aggregation of 100–1000 axons. Delay at synapses keeps velocity low in nerve nets and neuropiles.

The nerve impulse

A nerve impulse is a wave or zone of excitation that passes quickly down the length of the axon. Immediately ahead lies a region ready to be excited and immediately behind is a region temporarily powerless to conduct (refractory). The process is analogous to the movement of the actively burning region of a gunpowder fuse, in that the impulse occurs on adequate stimulus, regardless of its size. That is, if the size (voltage) and duration (milliseconds) of a stimulus is sufficient to excite an axon, the impulse is transmitted, and its character is not changed by a larger stimulus. A smaller voltage generally must be applied for longer time in order to be adequate. For example, a nerve of the crab *Cancer productus* was stimulated by 0.3 volt applied for 0.2 second or by 0.1 volt applied for 1.0 second. Each axon has a threshold of excitation and obeys the all-or-none principle. The analogy is not complete because the neuron has the power of quickly restoring the refractory region so that another impulse may follow. In general, variation in nerve messages is produced by relative frequency of impulses rather than their quality.

Models of nerve impulse

In spite of much description of neuronal behavior in electrical terms and the standard use of electrical stimuli, the actual conduction of an impulse is definitely not the same as conduction of electricity through a wire. Electricity flows at speeds approaching the velocity of light, 3×10^{10} cm/sec $= 3 \times 10^{8}$ m/sec, or about 10^{6} times the maximum velocity ever observed for a nerve impulse. However, an axon does have "cable properties," low longitudinal resistance and high membrane resistance. These help to explain the conduction of impulses or "electrotonic spread" of excitation. The membrane is, in fact, a capacitor, a device that holds a charge. The basic electrical definitions and units are: $I = E/R$, 1 ampere = 1 volt/1 ohm; $C = q/E$, 1 faraday = 1 coulomb/1 volt.

Many cells have membrane capacitance of $1 \mu f/cm^{2}$ and resistance of 1000 ohm/cm^{2} with conductance of cytoplasm 100 ohm^{-1}/cm^{2}.

The squid giant axon is so large that a microelectrode can be placed inside of it and paired with another on the outside to give a resting potential of the order of 0.06 volt, the interior negative. Furthermore, the axoplasm can be squeezed out and analyzed chemically, and even replaced with an artificial "cyto-plasm" of known composition. Analysis shows the following:

Loligo (Hodgkin, 1958)	K+	Na+	Cl⁻	organic anions
axoplasm	400	50	40	381
blood	20	440	560	—
sea water	10	460	540	0
Carcinus (Shaw, 1955)				
axoplasm	112	54	52	
blood	12	469	524	

The Nernst equation for resting potential

The potential across an artificial membrane is described approximately by the Nernst equation:

$$E = \frac{RT}{nf} \ln \frac{C_{ex}}{C_{in}}$$

E = potential in volts, R = 8.312 J/deg-mole, n = valency of ion, f = 96,500 coulombs/equiv = 96.5 coulombs/meq, T = degrees Kelvin, C_{in} and C_{ex} are interior and exterior concentrations of some crucial ion.

In the case of the squid axon, the crucial ion is K^+. For an axon at 18° C:

$$E_k = \frac{8.312 \times 291°}{1 \times 96.5} \ln 20/400 = 25.065 \ln 0.05 =$$
$$25.065(-2.9957) = -75.1 \, mV$$

A better prediction of the membrane potential is the weighted average of the equilibrium potentials of the major ions rather than K^+ alone. The resting conductances are G_{Na}= 0.0033 m mho/cm², G_K = 0.24, G_{Cl} = 0.30, indicating roughly equal permeability to K^+ and Cl^- and extremely low permeability to Na^+. The equilibrium potentials calculated from the Nernst equation can be substituted into the following equation to give a predicted membrane potential:

$$E_m = \frac{G_{Na}E_{Na} + G_K E_K + G_{Cl}E_{Cl}}{G_{Na} + G_K + G_{Cl}} =$$

$$\frac{[0.0033(55)] + [0.24(-75)] + [0.30(-67)]}{0.0033 + 0.24 + 0.30} =$$

$$\frac{0.1815 - 18.0 - 20.1}{0.5433} = -69.79 = -70\,mV$$

The role of the ions was elucidated with giant "cells" of the marine algae *Valonia* and *Halicystis*, then verified with squid axons after glass capillary microelectrodes were devised. These algae consist of a short stalk and a globose vesicle, 1 to 3 cm diameter, containing a large vacuole filled with "cell sap." The vesicle is not really a cell, but a coenocyte containing many nuclei and chloroplasts, produced by repeated cell division without cell wall formation. The Na^+ concentration of the medium was varied greatly without affecting the resting potential, while increasing K^+ in the medium reduced E all the way to zero when inner and outer concentrations were equal.

In the squid axon K^+ has been varied both inside and outside with predictable effect on E. While Cl^- concentration has some effect, it remains small. The maintenance of high internal K^+ is the work of a "sodium pump," which actively extrudes Na^+ and promotes K^+ entry in a coupled process. Most pumps that have been studied exchange 3 Na^+ for 2 K^+ at the expense of ATP hydrolysis. The positive charge on the surface of the axon is presumed due to the relative deficiency of Cl^- in the immediate vicinity, because of the very low Cl^- just inside, the strong gradient favoring inward movement. Permeability to Na^+ is small compared to K^+ and Cl^-.

The action potential

When an action potential occurs in a stimulated axon, the resting potential is not simply abolished, but actually reversed.

The reversal is very brief, about 1 msec, and then return to the resting potential occurs. The whole process takes about 3 msec, and appears as a "spike" on the screen of a cathode-ray oscilloscope. The upstroke and overshoot of the spike result from sudden increase in membrane permeability (Figure 11-1).

The conductances of Na^+ and K^+ increase enormously, the gates are opened, and the ions flow down their concentration gradients as if unimpeded: $G_{Na} = 120$ m mho/cm^2, $G_K = 36$. This predicts an altered membrane potential:

$$E_m = \frac{[120(55)] + [36(-75)] + [0.30(-67)]}{120 + 36 + 0.30} =$$

$$\frac{6600 - 2700 - 20.1}{156.30} = 24.82 = +25\,mV$$

The Hodgkin-Huxley model

The study of the nerve impulse was "one of the first and perhaps the most continuingly successful theoretical and experimental physical analyses of biological structure and function" (Cole, 1965). This statement is based on the Hodgkin-Huxley (1952) model of the squid axon, expressed by an equation of the form:

$$I_i = G_K n^4 (V - E_K) + G_{Na} m^3 h(V - E_{Na}) + G_L(V - E_L)$$

(G_L = non-specific conductances or leakage)

This predicts the instantaneous current flow (I_i) based on conductances (G) of the major ions and potentials (E) resulting from their unequal concentrations inside and outside the axon membrane. It is an elaboration of Ohm's (1827) law, $I = E/R$, as the reciprocal of resistance ($1/R$) is the same as conductance. Tests of the validity of this equation were performed through use of axial and external microelectrodes, with membrane potential restricted electronically by a "voltage clamp" apparatus, which held Vm constant.

Maximum current flow under voltage clamp is about 10 m

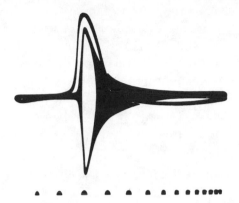

Figure 11-1. Action potential of squid giant axon as shown by cathode-ray oscilloscope. Increase of membrane conductance (band), and change of membrane potential (line) during impulse. 1 msec marks. (From Cole, 1965).

amp/cm^2. Calculated currents agree very well with observed, and one result was a calculated propagation velocity of 18.8 m/sec, 88.7 percent of the observed 21.2 m/sec.

Exceptions to the model axon

Although the squid giant axon and the model derived from its experimental use have been highly successful, it is not correct to assume that every axon behaves exactly like it. Neurons may activate other neurons without displaying a typical action potential, but merely a slight depolarization with no overshoot or reversal of potential. This is seen in barnacle photoreceptors (4). The median ocellus is not an image-producing eye, but a simple light-sensitive organ consisting of only four cells. The ocellus was isolated from *Balanus nubilis*, along with the principal ganglia and the ocellar nerve connecting them. In response to light, the cell body of an ocellus depolarizes, and the wave of depolarization spreads down the axon to terminals in the supra-esophageal ganglion, where it can be recorded with an

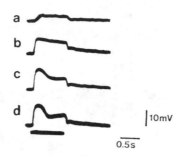

Figure 11–2. Intracellular depolarization of axon terminal branch of ocellar nerve of barnacle *Balanus nubilis* in response to varied light stimuli. Intensity of illumination increasing from a to d, 4.9×10^3 to 1.8×10^6 quanta/sec per μm^2. (From Ozawa *et al.*, 1976).

intracellular microelectrode (Figure 11-2). The steady-state depolarization 10 mm down the axon is about one-third that seen in the cell body, i.e., about -20 mV rather than -60 mV. The function of the ganglion is to somehow "invert" the signal, so that the barnacle ignores a sustained light and responds only to a sudden decrease in intensity by withdrawing into the shell— the "shadow reflex." Connective neurons in the ganglion respond to stimulation by the ocellar nerve with typical spikes, in the same manner as the squid axon.

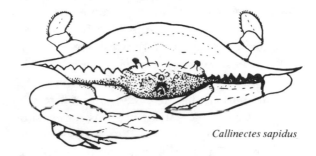

Callinectes sapidus

CHAPTER 12 Hormonal effects

Female blue crabs, in the course of a few weeks, undergo their final juvenile molt, mate, and develop ovaries. As the female extrudes eggs they are fertilized, and embryonic development begins while the eggs are held in the space provided by loosening of the formerly tightly flexed abdomen. Zoea larvae hatch out, and a new generation of crabs begins a set pattern of growth, molting, and maturation.

Growth and reproduction are controlled by hormones, chemical messengers produced by special glandular tissues under the influence of seasonal or other appropriate environmental stimuli. Hormone concentrations in the blood gradually build up to the level necessary to activate a target organ such as the external epithelium or the gonads. Hormonal effects are known in many marine invertebrates, but only in decapod crustaceans are the hormones sufficiently characterized to discuss them in chemical terms.

Crustacean molting

The order Decapoda is the largest order of crustaceans (8500 species) and contains the largest species, such as the edible lobsters and crabs. The blue crab, *Callinectes sapidus*, is a member of the Portunidae, the swimming crabs. All decapods have three pairs of thoracic appendages modified into maxillipeds, and in the portunids, the first of the five pairs of walking legs are chelipeds and the last are swimming legs. The integument, consisting of protein, chitin, and calcium carbonate, influences "every phase of crustacean biology" (7). Molting is a normal part of the life cycle, not an interruption of it.

The Crustacea consist of one large sub-class, Malacostraca, which contains the familiar conspicuous decapods, isopods, amphipods, and euphausids. There are also seven lesser sub-classes, including the branchiopods, ostracods, copepods, barnacles, and three sub-classes with very few species.

Crustecdysone

In the head region of more than fifty species of malacostracans, a gland located in the antennary or second maxilliary segment produces a molting hormone. Removal of this gland stops the molting cycle, and its implantation brings on the pre-molt condition (3). The molting hormone, crustecdysone, is a steroid, with the same basic structure as cholesterol, Vitamin D, the bile acids, adrenal cortical hormones, and sex hormones of mammals, the saponins and the cardiac glycosides. The structure of crustecdysone (Figure 12–1) differs only in the addition of a single hydroxyl group from that of ecdysterone, the molting hormone of insects, which was first deduced in 1965 and synthesized in a twenty-two step process from stigmasterol in 1967.

Two related steroids called callinecdysones A and B occur in addition to crustecdysone in *Callinectes sapidus* (1). Three stages of molt were examined for these compounds: The early pre-molt contained only A, 5 μg/kg tissue, the late pre-molt or "peeler" had 20 μg of A and 4 μg of crustecdysone per kg, and the newly molted female crabs, "soft shells," had a very large quantity of crustecdysone, 280 μg/kg, and a smaller amount of

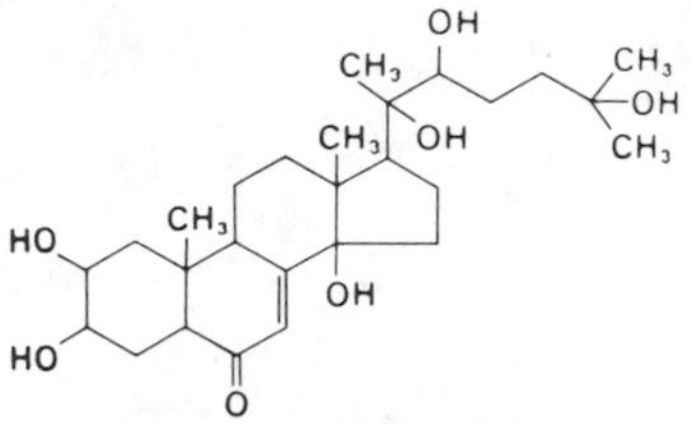

Figure 12-1. Structure of 20-hydroxy-ecdysone (crustecdysone), the molting hormone of crustaceans. The 20 carbon is just above the cyclopentane ring.

B, $24 \mu g/kg$. The spectra of A and B were identical to those of inokosterone and makisterone respectively, compounds originally isolated from plants. The high concentration of crustecdysone in crabs that had already molted suggests that the hormone affects not only shedding but hardening of the cuticle as well. Actually, the use of words like "very large quantity" and "high concentration" are relative, since hormone concentrations are always very small by comparison with other metabolites. If 1 kg of crabs contains 800 ml water, then crustecdysone, with a molecular weight of 479.6, reached a maximum concentration of $7.3 \times 10^{-7}M$. The concentration in "peelers" was $1.0 \times 10^{-8}M$. Thus the change in concentration was seventyfold between stages in the molting cycle. To approach the understanding of endocrinology now already reached for some vertebrates, the blood concentrations of all three molting hormones and one or more molt inhibiting hormones should be measured each day of the cycle. No laboratory has had the mission or resources to do this yet. The starting material for each analysis would consist of about a hundred large crabs, all in exact phase with one another. This almost presumes laboratory rearing from eggs of a single spawning. Work with larval crustaceans has progressed very well, but rearing to adult size would require much more aquarium space, labor, and time.

Y-organ and X-organ

The difficulty of establishing endocrine function in a crustacean tissue lies in the very small size of presumed organs, lack of detailed anatomy of many species, and the secretion of hormones by neurons rather than typical glandular cells. These factors mitigate against the standard proof of function, which consists of surgical extirpation, followed by restoration of endocrine activity by means of implantation, and maintenance of normal balance by chemically pure active principle extracted from the organ. Nevertheless, the gland producing crustecdysone has been found in several species. In some crabs this pair of glands, called Y-organs, degenerates after the last molt. The activity of the molting gland is controlled by a neurosecretory center in the eyestalk, which produces an inhibitory hormone. This center has been called the X-organ (4).

The neurosecretory hormone that inhibits Y-organ activity has not been isolated and characterized yet. However, two other eyestalk hormones have been identified, and they are polypeptides. The blanching hormone or "red-pigment concentrating hormone" is an octapeptide of known amino acid sequence (Figure 12-2). It was isolated from eyestalks of a prawn, *Pandalus borealis*, and the structure was determined by analysis and confirmed by synthesis in 1972 (2). Through either extraordinary coincidence or some principle of molecular design, at least two neurosecreted hormones of vertebrates also have the amino terminal pyroglutamic acid and the carboxyl terminal amide group. The "retinal-pigment-light-adapting hormone" is a larger peptide of eighteen residues of twelve different amino acids, sequence not yet determined. Therefore, it is likely that the molt-inhibiting hormone is also a polypeptide.

Effects of de-stalking

Since it is very difficult to find the precise location of the neurosecretory cells responsible for synthesis of each of the

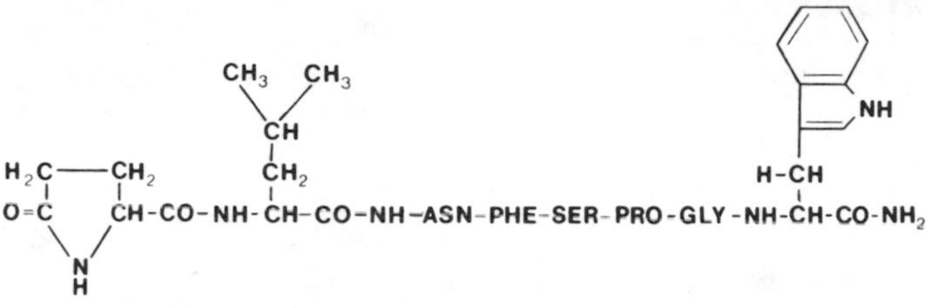

Figure 12-2. Structure of the blanching hormone from the eyestalk of *Pandalus borealis*. It has an amino-terminal pyroglutamic acid and a carboxyl-terminal amide group. (From Fernlund and Josefsson, 1972.)

several hormones, a common experimental practice is to remove the entire eyestalk and study some activity of one of the target tissues or organs. After "de-stalking," a crab (*Cancer irroratus*) was injected with ³H-uridine, a precursor of RNA, and the amount of tracer in sections of Y-organ was determined by radioautography (6). In this case, the paired Y-organs were 1–2 mm pale yellow structures in the cephalothorax, anterior to the branchial chamber, just posterior to the eyestalks, and directly applied to the hypodermis of the ventral carapace. The most profound effects were seen after 3.5 days (Figure 2-3). There was a fourfold increase in RNA synthesis, as measured by incorporation of the tritium atoms and appearing as silver grains on the photographic emulsion applied to the tissue section. At the same time the number of nuclei per unit area had diminished to a minimum, indicating increased cytoplasmic volume or cellular hypertrophy. This experiment does not prove that crustecdysone production increases upon release from inhibition by the neurosecretory complex, but it does show that some kind of increased synthetic activity occurs in the Y-organs during a critical period and is followed by a gradual return toward the control state.

Eyestalk extracts also exert effects on the concentration of glucose in blood, retinal pigment migration, salt and water balance, ovarian development, and color change.

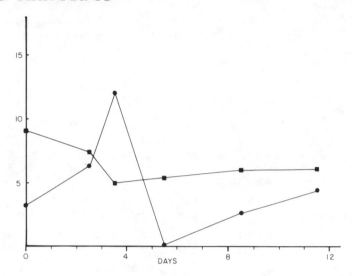

Figure 12–3. RNA synthesis in the Y-organ of *Cancer irroratus*, measured by incorporation of ^{3}H-uridine, reaches a maximum 3.5 days after removal of eyestalk. Volume of cells in the gland reaches a maximum at the same time, as indicated by minimum number of nuclei observed in standard area of tissue.

Control of color change

Many decapod crustaceans can change their body color by means of chromatophores, hypodermal pigment cells with branched processes. The animal becomes lighter in color when the pigment granules move from the branches to a tightly-packed central depot. As a protective measure, the animal usually changes color to match its background more closely.

In some species, the pigment-dispersing hormone is mediated by adenosine 3', 5' cyclic monophosphate (cAMP), in accordance with the hypothesis that peptide hormones in general act on chromatophores by stimulating production of a "second messenger," such as cAMP, within the target cells (5). In the prawn *Palaemonetes pugio*, recent evidence supports the idea that the second messenger for red-pigment concentrating hormone (Figure 12-2) is calcium ion. Concentration of the granules in chromatophores of several tissues was inhibited by replacing Ca^{++} with La^{++} in the medium. An ionophore is a compound of moderate molecular weight, 200 to 2,000, that contains oxygen atoms arranged to form a cavity into which a cation fits closely, forming a lipid-soluble complex favoring transport across membranes. A Ca^{++}-transporting ionophore produced pigment concentration in *Palaemonetes* as rapidly as the blanching hormone, but in Ca^{++}-free medium the ionophore was ineffective.

Improvement of our understanding of hormonal control can be expected as more hormones are chemically identified and applied to specific target cells in measured doses.

References

Chapter 1: Osmotic balance

1. Allen, J.A. and M.R. Garrett. 1972. Studies on taurine in the euryhaline bivalve *Mya arenaria. Comp. Biochem. Physiol.* 41:307-17.
2. Avens, A.C. and M.A. Sleigh. 1965. Osmotic balance in gastropod mollusks. I. Some marine and littoral gastropods. *Comp. Biochem. Physiol.* 16:121-41.
3. Binns, R. 1969. The physiology of the antennal gland of *Carcinus maenas* L. II. Urine production rates. *J. Exptl. Biol.* 51:11-16.
4. Bishop, S.H. 1976. Nitrogen metabolism and excretion: regulation of intracellular amino acid concentrations. In *Estuarine Processes*, vol. 1, ed. M. Wiley, pp. 414-31. New York: Academic Press.
5. Bricteux-Grégoire, S.; Gh. Duchâteau-Bosson; Ch. Jeuniaux; and M. Florkin. 1964. Constituants osmotiquement actifs des muscles adducteurs de *Gryphaea angulata* adaptée à l'eau de mer ou à l'eau saumâtre. *Arch. Int. Physiol. Biochim.* 72:835-42.
6. Freel, R.W.; S.G. Medler; and Mary E. Clark. 1973. Solute adjustments in the coelomic fluid and muscle fibers of a euryhaline polychaete, *Neanthes succinea*, adapted to various salinities. *Biol. Bull.* 144:289-303.
7. Freeman, R.F.H. and T.J. Shuttleworth. 1977. Distribution of water in *Arenicola marina* L. equilibrated to diluted sea water. *J. Mar. Biol. Assoc. U.K.* 57:501-19.
8. Gilles, R. 1972. Osmoregulation in three molluscs: *Acanthochitona discrepans* Brown, *Glycymeris glycymeris* L. and *Mytilus edulis* L. *Biol. Bull.* 142:25-35.

9. Haberfield, Eve C.; L.W. Haas; and C.S. Hammen. 1975. Early ammonia release by a polychaete *Nereis virens* and a crab *Carcinus maenas* in diluted sea water. *Comp. Biochem. Physiol.* 52A:501-3.
10. Kinne, O. 1971. Salinity: invertebrates. In *Marine ecology*, vol. 1, part 1, ed. O. Kinne, pp. 821-995. New York: Wiley-Interscience.
11. Krogh, A. 1939. *Osmotic regulation in aquatic animals.* New York: Dover.
12. Lange, R. 1972. Some recent work on osmotic, ionic and volume regulation in marine animals. *Oceanogr. Mar. Biol. Ann. Rev.* 10:97-136.
13. Potts, W.T.W. 1968. Osmotic and ionic regulation. *Ann. Rev. Physiol.* 30:73-104.
14. Potts, W.T.W. and G. Parry. 1964. *Osmotic and ionic regulation in aminals.* New York: Macmillan.
15. Schoffeniels, E. 1964. Effect of inorganic ions on the activity of L-glutamic acid dehydrogenase. *Life Sci.* 3:845-50.
16. Shaw, J. 1955. Ionic regulation in the muscle fibers of *Carcinus maenas*. I. The electrolyte composition of single fibers. *J. Exptl. Biol.* 32:383-96.
17. Shaw, J. 1958. Osmoregulation in the muscle fibers of *Carcinus maenas*. *J. Exptl. Biol.* 35:920-29.
18. Smith, R.I. 1976. Exchanges of sodium and chloride at low salinities by *Nereis diversicolor* (Annelida, Polychaeta). *Biol. Bull.* 151:587-600.
19. Spaargaren, D.H. 1974. A study of the adaptation of marine organisms to changing salinities with special reference to the shore crab *Carcinus maenas* L. *Comp. Biochem. Physiol.* 47A:499-512.

Chapter 2: Gaseous exchange

1. Brown, A.C. and G. Brengelmann. 1965. Energy metabolism. In *Physiology and biophysics*, ed. T.C. Ruch and H.D. Patton, pp. 1030-49. Philadelphia: Saunders.
2. Galtsoff, P.S. 1964. The American oyster, *Crassostrea virginica. Fishery Bull. 64.* Washington: U.S. Govt. Printing Office.
3. Grainger, J.N. 1968. The relation between heat production, oxygen consumption, and temperature in some poikilotherms. In *Quantitative biology of metabolism*, ed. A. Locker, pp. 86-89. New York: Springer-Verlag.
4. Hammen, C.S. 1972. Lactate oxidation in the upper-shore barnacle, *Chthamalus depressus* (Poli.). *Comp. Biochem. Physiol.* 43A:435-41.
5. Hammen, C.S.; D.P. Hanlon; and S.C. Lum. 1962. Oxidative metabolism of *Lingula*. *Comp. Biochem. Physiol.* 5:185-91.
6. Hammen, C.S. and S.C. Lum. 1964. Carbon dioxide fixation in marine invertebrates: quantitative relations. *Nature* 201:416-17.
7. Kalle, K. 1972. Dissolved gases. In *Marine Ecology*, vol. 1, part 3, ed. O. Kinne, pp. 1451-57. New York: Wiley-Interscience.

8. Krogh, A. 1941. *The comparative physiology of respiratory mechanisms.* New York: Dover.
9. Umbreit, W.W.; R.H. Burris; and J.F. Stauffer. 1972. *Manometric and biochemical techniques.* 5th ed. Minneapolis: Burgess.
10. Van Winkle, W. and C. Mangum. 1975. Oxyconformers and oxyregulators: a quantitative index. *J. Exp. Mar. Biol. Ecol.* 17:103–10.
11. Warburg, O.H. and G. Krippahl. 1959. Weiterentwicklung der manometrischen Methoden. In *New methods of cell physiology,* ed. O.H. Warburg, pp. 374–77. New York: Wiley.

Chapter 3: Oxygen transport

1. Bohr, C.; K. Hasselbalch; and A. Krogh. 1904. Ueber einem in biologischer Beziehung wichtigen Einfluss, den die Kohlensäurespannung des Blutes auf dessen Sauerstoffbindung übt. *Skand. Arch. Physiol.* 16:402–12.
2. Garlick. R.L. and R.C. Terwilliger. 1977. Structure and oxygen equilibrium of hemoglobin and myoglobin from the Pacific lugworm *Abarenicola pacifica. Comp. Biochem. Physiol.* 57B:177–84.
3. Krogh, A. 1941. *The comparative physiology of respiratory mechanisms.* New York: Dover.
4. Mangum, C.P. 1976. The oxygenation of hemoglobin in lugworms. *Physiol. Zool.* 49:85–99.
5. Manwell, C. 1960. Comparative physiology: blood pigments. *Ann. Rev. Physiol.* 22:191–244.
6. Prosser, C.L., ed. 1973. *Comparative animal physiology.* 3rd ed. Philadelphia: Saunders.
7. Redmond, J.R. 1962. The respiratory characteristics of chiton hemocyanins. *Physiol. Zool.* 35:304–13.
8. Redmond, J.R. 1968. The respiratory function of hemocyanin. In *Physiology and biochemistry of haemocyanins,* ed. F. Ghiretti. New York: Academic Press.
9. Terwilliger, R.C. and K.R.H. Read. 1969. The radular muscle myoglobins of the amphineuran mollusk, *Acanthopleura granulata* Gmelin. *Comp. Biochem. Physiol.* 29:551–60.
10. Waxman, L. 1971. The hemoglobin of *Arenicola cristata. J. Biol. Chem.* 246:7318–27.
11. Young, R.E. 1972. The physiological ecology of haemocyanin in some selected crabs. 1. The characteristics of haemocyanin in a tropical population of the blue crab *Callinectes sapidus* Rathbun. *J. Exp. Mar. Biol. Ecol.* 10:183–92.

Chapter 4: Digestion

1. Baldwin, E. 1964. *An introduction to comparative biochemistry.* 4th ed. Cambridge: University Press.
2. Doyle, J. 1966. Studies on the chemical nature of the crystalline

style. In *Some contemporary studies in marine science*, ed. H. Barnes. New York: Hafner.

3. Enzyme Commission. 1973. *Enzyme nomenclature*. Recommendations of the International Union of Biochemistry. rev. ed. New York: Elsevier.

4. Kristensen, J.H. 1972. Carbohydrases of some marine invertebrates with notes on their food and on the natural occurrence of the carbohydrates studied. *Mar. Biol.* 14:130-42.

5. Lineweaver, H. and D. Burk. 1934. Determination of enzyme dissociation constants. *J. Amer. Chem. Soc.* 56:658-66.

6. Owen, G. 1974. Feeding and digestion in the Bivalvia. *Adv. Comp. Physiol. Biochem.* 5:1-35.

7. Pandian, T.J. 1975. Mechanisms of heterotrophy. In *Marine Ecology*, vol. 2, part 1, ed. O. Kinne. New York: Wiley.

8. Patton, J.S. and J.G. Quinn. 1973. Studies on the digestive lipase of the surf clam *Spisula solidissima. Mar. Biol.* 21:59-69.

Chapter 5: Intermediary Metabolism

1. Berg, W.E. and D.H. Mertes. 1970. Rates of synthesis and degradation of protein in the sea urchin embryo. *Exptl. Cell Res.* 60:218-24.

2. Chen, C.H. and A.L. Lehninger. 1973. Respiration and phosphorylation by mitochondria from the hepatopancreas of the blue crab, *Callinectes sapidus. Arch. Biochem. Biophys.* 154:449-59.

3. Corrigan, J. 1969. D-amino acids in animals. *Science* 164:142-49.

4. Davis, R.H.; K.L. Sisco; and F.R. Sharp. 1977. Protein synthetic activity in nervous tissue from different anatomical regions of lobster. *Comp. Biochem. Physiol.* 57:65-71.

5. Ellington, W.R. and G.L. Long. 1978. Purification and characterization of a highly unusual tetrameric D-lactate dehydrogenase from the muscle of the giant barnacle, *Balanus nubilis* Darwin. *Arch. Biochem. Biophys.* 186:265-74.

6. Florkin, M. 1966. *A molecular approach to phylogeny*. New York: Elsevier.

7. Florkin, M. and S. Bricteux-Grégoire. 1972. Nitrogen metabolism in mollusks. In *Chemical zoology*, vol. 7, eds. M. Florkin and B.T. Scheer, pp. 301-48. New York: Academic Press.

8. Hammen, C.S. 1968. Aminotransferase activities and amino acid excretion of bivalve mollusks and brachiopods. *Comp. Biochem. Physiol.* 26:697-705.

9. Hammen, C.S. 1977. Brachiopod metabolism and enzymes. *Amer. Zool.* 17:141-47.

10. Hammen, C.S.; D.P. Hanlon; and S.C. Lum. 1962. Oxidative metabolism of *Lingula. Comp. Biochem. Physiol.* 5:185-91.

11. Hanlon, D.P.; L. DeVore; M.C. Kincaid; A. Jones; and L. Lane. 1971. Comparative enzymology of molluscan enolases. *Comp. Biochem. Physiol.* 39:383-93.

12. Jodrey, L.H. and K.M. Wilbur. 1955. Studies on shell formation. IV. The respiratory metabolism of the oyster mantle. *Biol. Bull.* 108: 346-58.
13. Kittredge, J.S.; E. Roberts; and D.G. Simonsen. 1962. The occurrence of free 2-aminoethylphosphonic acid in the sea aneomone, *Anthopleura elegantissima. Biochem.* 1:624-28.
14. Long, G.L. 1976. The stereospecific distribution and evolutionary significance of invertebrate lactate dehydrogenases. *Comp. Biochem. Physiol.* 55:77-83.
15. Quin, L.D. and F.A. Shelburne. 1969. An examination of marine animals for the presence of carbon-bound phosphorus. *J. Mar. Res.* 27:73-84.
16. Thabrew, M.I.; P.C. Poat; and K.A. Munday. 1971. Carbohydrate metabolism in *Carcinus maenas* gill tissue. *Comp. Biochem. Physiol.* 40:531-41.
17. Yamanaka, T.; H. Mizushima; and K. Okunuki. 1964. Cytochrome C's purified from marine invertebrates. *Biochem. Biophys. Acta.* 81:223-28.
18. deZwaan, A. and D.I. Zandee. 1972. The utilization of glycogen and accumulation of some intermediates during anaerobiosis in *Mytilus edulis* L. *Comp. Biochem. Physiol.* 43:47-54.

Chapter 6: Total metabolism

1. Becker, W. and I. Lamprecht. 1977. Mikrokalorimetrische Untersuchungen zum Wirt-Parasit-Verhältnis zwischen *Biomphalaria glabrata* und *Schistosoma mansoni. Z. f. Parasitenkunde.* 53:297-305.
2. Grainger, J.N.R. 1968. The relation between heat production, oxygen consumption and temperature in some poikilotherms. In *Quantitative biology of metabolism.* ed. A. Locker, pp. 86-89. New York: Springer-Verlag.
3. Hammen, C.S. 1979. Metabolic rates of marine bivalve mollusks determined by calorimetry. *Comp. Biochem. Physiol.* 62A:955-59.
4. Hill, A.V. 1911. The total energy exchange of intact cold-blooded animals at rest. *J. Physiol.* 43:379-94.
5. Kleiber, M. 1961. *The fire of life: an introduction to animal energetics.* New York: Wiley.
6. Lamprecht, I. 1976. Application of calorimetry to the evaluation of metabolic data for whole organisms. *Biochem. Soc. Trans.* 4:565-69.
7. Lavoisier, A.L. and P. Laplace. 1780. Memoire sur la chaleur. Mem. de l'Acad. des Sciences. Paris. Translation in: M.L. Gabriel and S. Fogel. 1955. *Great experiments in biology.* Englewood Cliffs, N.J.: Prentice-Hall.
8. Meyerhof, O. 1911. Untersuchungen über die Wärmetönung der vitalen Oxydationsvorgänge in Eiern. II. *Biochem. Zeit.* 35:280-315.
9. Pamatmat, M.M. 1978. Oxygen uptake and heat production in a

metabolic conformer (*Littorina irrorata*) and a metabolic regulator (*Uca pugnax*). *Mar. Biol.* 48:317-25.
10. Spaargaren, D.H. 1975. Heat production of the shore-crab *Carcinus maenas* (L.) and its relation to osmotic stress. In *Proc. Ninth Europ. Mar. Biol. Symp*, ed. H. Barnes, pp. 475-82. Aberdeen, Scotland: Aberdeen University Press.
11. deZwaan, A. 1977. Anaerobic energy metabolism in bivalve molluscs. *Oceanogr. Mar. Biol. Ann. Rev.* 15:103-87.

Chapter 7: Nitrogen excretion

1. Campbell, J.W. 1973. Nitrogen excretion. In *Comparative animal physiology*, 3rd ed., ed. C.L. Prosser, pp. 279-316. Philadelphia: Saunders.
2. Florkin, M. and S. Bricteux-Grégoire. 1972. Nitrogen metabolism in mollusks. In *Chemical Zoology*, vol. 7, ed. M. Florkin and B.T. Scheer, pp. 301-48. New York: Academic Press.
3. Haberfield, E.C.; L.W. Haas; and C.S. Hammen. 1975. Early ammonia release by a polychaete *Nereis virens* and a crab *Carcinus maenas* in diluted sea water. *Comp. Biochem. Physiol.* 52:501-3.
4. Hammen, C.S. 1968. Aminotransferase activities and amino acid excretion of bivalve mollusks and brachiopods. *Comp. Biochem. Physiol.* 26: 697-705.
5. Hammen, C.S.; H.F. Miller; and W.H. Geer. 1966. Nitrogen excretion of *Crassostrea virginica. Comp. Biochem. Physiol.* 17:1199-1200.
6. Lewis, J.B. 1967. Nitrogenous excretion in the tropical sea urchin *Diadema antillarum* Philippi. *Biol. Bull.* 132:34-37.
7. Lum, S.C. and C.S. Hammen. 1964. Ammonia excretion of *Lingula. Comp. Biochem. Physiol.* 12:185-90.
8. Razet, P. and C. Retière. 1967. Recherche des enzymes de la chaine de l'uricolyse chez les Annélidès polychètes. *Comptes rendus. Acad. Sci. Paris.* 264D:356-59.
9. Riley, J.P. and D.A. Segar. 1970. The seasonal variation of the free and combined dissolved amino acids in the Irish Sea. *J. Mar. Biol. Ass.* 50: 713-20.
10. Stephens, G.C. and R.A. Schinske. 1961. Uptake of amino acids by marine invertebrates. *Limnol. Oceanog.* 6:175-81.
11. Wood, L. 1965. Determination of free amino acids in sea water. In *Automation in analytical chemistry*, ed. L.T. Skeggs, Jr. New York: Mediad.
12. Yamin, M.; D.R. Crawford; M. Minor; and S.H. Bishop. 1977. Arginine and urea biosynthesis in the lugworm, *Arenicola cristata. Comp. Biochem. Physiol.* 57:223-25.

Chapter 8: Shell formation

1. Chapheau, M. 1932. Recherches sur le métabolisme cellulaire de quelques invertébrés marins. *Bull. Sta. Biol. Arcachon.* 29:85-152.

2. Florkin, M. 1971. The present state of molecular paleontology. In *Chemical evolution and the origin of life*, ed. R. Buvet and C. Ponnamperuma, pp. 10–26. New York: American Elsevier.
3. Hammen, C.S. 1969. Lactate and succinate oxidoreductases in marine invertebrates. *Mar. Biol.* 4:233–38.
4. Harriss, R.C. 1965. Trace element distribution in molluscan skeletal material. I. Magnesium, iron, manganese, and strontium. *Bull. Mar. Sci.* 15:265–73.
5. Lowenstam, H.A. and D.P. Abbott. 1975. Vaterite: a mineralization product of the hard tissues of a marine organism (Ascidiacea). *Science.* 188:363–65.
6. Mori, K. 1968. Changes of oxygen consumption and respiratory quotient in the tissues of oysters during the stages of sexual maturation and spawning. *Tohoku J. Agr. Res.* 19:136–43.
7. Percy, J.A.; F.A. Aldrich; and T.R. Marcus. 1971. Influence of environmental factors on respiration of excised tissues of American oysters, *Crassostrea virginica* Gmelin. *Canad. J. Zool.* 49:353–60.
8. Rudwick, M.J.S. 1970. *Living and fossil brachiopods.* London: Hutchinson.
9. Watabe, N.; D.G. Sharp; and K.M. Wilbur. 1958. Studies on shell formation. VIII. Electron microscopy of crystal growth on the nacreous layer of the oyster *Crassostrea virginica. J. Biophys. Biochem. Cytol.* 4:281–86.
10. Wilbur, K.M. 1972. Shell formation in mollusks. In *Chemical zoology*, vol. 7, eds. M. Florkin and B.T. Scheer, pp. 103–45. New York: Academic Press.

Chapter 9: Bioluminescence

1. Clayton, R.K. 1971. *Light and living matter: a guide to the study of photobiology.* vol. 2. New York: McGraw-Hill.
2. Cormier, J.J.; J. Lee; and J.E. Wampler. 1975. Bioluminescence: recent advances. *Ann. Rev. Biochem.* 44:255–72.
3. Florkin, M. and S. Bricteux-Grégoire. 1972. Nitrogen metabolism in mollusks. In *Chemical zoology*, vol. 7, eds. M. Florkin and B.T. Scheer, pp. 301–48. New York: Academic Press.
4. Fridovich, I. 1975. Oxygen: boon and bane. *Amer. Scientist.* 63:54–59.
5. Harvey, E.N. 1952. *Bioluminescence.* New York: Academic Press.
6. Johnson, F.H. and Y. Haneda, eds. 1966. *Bioluminescence in progress.* Princeton, N.J.: Princeton University Press.
7. Johnson, F.H. and O. Shimomura. 1968. The chemistry of luminescence in coelenterates. In *Chemical zoology*, vol. 7, eds. M. Florkin and B.T. Scheer, pp. 233–61. New York: Academic Press.
8. Johnson, F.H. and O. Shimomura. 1975. Bacterial and other luciferins. *Bioscience* 25:718–23.

Chapter 10: Ciliary activity and muscular contraction

1. Ashley, C.C. and E.B. Ridgway. 1970. On the relationships between membrane potential, calcium transient and tension in single barnacle muscle fibers. *J. Physiol.* 209:105–30.
2. Brokaw, C.J. 1967. Adenosine triphosphate usage by flagella. *Science* 156:76–78.
3. Cohen, C. and A.G. Szent-Gyorgyi. 1971. Assembly of myosin filaments and the structure of molluscan catch muscles. In *Contractility of muscle cells and related processes,* ed. R.J. Podolsky, pp. 23–36. Englewood Cliffs, N.J.: Prentice-Hall.
4. Hill, A.V. 1970. *First and last experiments in muscle mechanics.* Cambridge: University Press.
5. Hoyle, G. and T. Smyth, Jr. 1963. Giant muscle fibers in a barnacle, *Balanus nubilis* Darwin. *Science* 139:49–50.
6. Mendelson, M. 1969. Electrical and mechanical characteristics of a very fast lobster muscle. *J. Cell Biol.* 42:548–63.
7. Millman, B.M. 1964. Contraction in the opaque part of the adductor muscle of the oyster (*Crassostrea angulata*). *J. Physiol.* 173:238–62.
8. Pandian, T.J. 1975. Mechanisms of heterotrophy. In *Marine Ecology,* vol. 2, part 1, ed. O. Kinne, pp. 61–249. New York: Wiley.
9. Satir, P. 1974. The present status of the sliding microtubule model of ciliary motion. In *Cilia and flagella,* ed. M.A. Sleigh, pp. 131–42. New York: Academic Press.
10. Sleigh, M.A. ed. 1974. *Cilia and flagella.* New York: Academic Press.
11. Stephens, R.E. 1971. Biochemistry of cilia and flagella. In *Contractility of muscle cells and related processes,* ed. R.J. Podolsky, pp. 55–66. Englewood Cliffs, N.J.: Prentice-Hall.
12. Watts, R.L. 1971. Genes, chromosomes and molecular evolution. In *Biochemical evolution and the origin of life,* ed. E. Schoffeniels, pp. 14–42. Amsterdam: North-Holland.

Chapter 11: Nervous conduction

1. Cole, K.S. 1965. Theory, experiment, and the nerve impulse. In *Theoretical and mathematical biology,* eds. T.H. Waterman and H.J. Morowitz, pp. 136–71. New York: Blaisdell.
2. Eckert. R. 1978. *Animal physiology.* San Francisco: W.H. Freeman.
3. Hodgkin, A.L. and A.F. Huxley. 1952. A quantitative description of membrane current and its application to conduction and excitation in nerve. *J. Physiol.* 117:500–544.
4. Ozawa, S.; S. Hagiwara; K. Nicolaysen; and A.E. Stuart. 1976. Signal transmission from photoreceptors to ganglion cells in the visual system of the giant barnacle. *Cold Spring Harbor Symposium on Quantitative Biology.* 40:563–70.

Chapter 12: Hormonal effects

1. Faux, A.; D.H.S. Horn; E.J. Middleton; H.M. Fales; and M.E. Lowe. 1969. Molting hormones of a crab during ecdysis. *Chem. Comm.* 175–76.
2. Fernlund, P. and L. Josefsson. 1972. Crustacean color-change hormone: amino acid sequence and chemical synthesis. *Science.* 177:173–75.
3. Gilbert, L.I. 1963. Hormones controlling reproduction and molting in invertebrates. In *Comparative endocrinology*, eds. U.S. von Euler and H. Heller, chap. 14. New York: Academic Press.
4. Highnam, K.C. and L. Hill. 1977. *The comparative endocrinology of the invertebrates.* 2nd ed. London: Arnold.
5. Lambert, D.T. and M. Fingerman. 1979. Evidence implicating calcium as the second messenger for red pigment-concentrating hormone in the prawn *Palaemonetes pugio. Physiol. Zool.* 52:497–508.
6. Simione, F.P. and D.L. Hoffman. 1975. Some effects of eyestalk removal on the Y-organs of *Cancer irroratus* Say. *Biol. Bull.* 148: 440–47.
7. Waterman, T.H. and F.A. Chace. 1960. General crustacean biology. In *Physiology of Crustacea*, ed. T.H. Waterman, chap. 1. New York: Academic Press.

Index

Library of Congress Cataloging in Publication Data

Hammen, Carl Schlee.
  Marine invertebrates.

  Bibliography: p.
  Includes index.
  1. Marine invertebrates—Physiology.
  2. Physiology, Comparative. I. Title. [DNLM:
  1. Invertebrates—Physiology. 2. Marine biology.
  3. Physiology, Comparative. QL364 H224m]
  QL364.H35   592.092   80–51505
  ISBN 0-87451-188-7